YOUR SAFETY
IS THE #1 PRIORITY.

The purpose of this book is to advance the cause of fire safety.

Our country has the largest per capita fire mortality rate in the world. If we could only invest some time and a little money into intelligent fire-safety preplanning, we could have the lowest rate of fire and fire mortality in the world.

This is a simple, straightforward guidebook, for I believe that clever phrases and fancy technological catchwords will not save a single life—but sound, basic commonsense information will. It is designed section by section, building environment by building environment, to increase your awareness of fire problems and to protect yourself and your loved ones from the perils of fire and smoke.

DENNIS SMITH'S
FIRE SAFETY BOOK

DENNIS SMITH'S FIRE SAFETY BOOK:

Everything You Need to Know To Save Your Life

BANTAM BOOKS
TORONTO · NEW YORK · LONDON · SYDNEY

DENNIS SMITH'S FIRE SAFETY BOOK:
EVERYTHING YOU NEED TO KNOW
TO SAVE YOUR LIFE
A Bantam Book / March 1983

ISBN 0-553-23061-1

Published simultaneously in the United States and Canada

Bantam Books are published by Bantam Books, Inc. Its trademark, consisting of the words "Bantam Books" and the por-
trayal of a rooster, is Registered in U.S. Patent and Trademark
Office and in other countries, Marca Registrada. Bantam
Books, Inc., 666 Fifth Avenue, New York, New York 10103.

PRINTED IN THE UNITED STATES OF AMERICA

O   0 9 8 7 6 5 4 3 2 1

CONTENTS

INTRODUCTION

In the more than seventeen years I have worked as a New York City fire fighter, I have been inside hundreds of burning buildings, many buildings where everyone escaped. There were others where people did not get out, where they were trapped and overcome by heat or flames or both. Fortunately, we saved some of these people. Tragically, there were those we could not save, and they perished.

It might be justifiably asked: Why do we have such a high rate of fire and fire mortality in the United States of America? Why are we indifferent to the confronting possibility of our own destruction? The answers, I think, lie in the very structures of the American personality. We are still the adventurers, the men and women moving around in Conestoga wagons, seeking the fruits of America's bounty, though the wagons are as likely now to be Sevilles and the shelter not a pitched tent but a swank high-rise hotel. There is no longer any land to be discovered, yet we still search relentlessly for the fortunate future, the job, the opportunity, the connection that will make us one of America's elect rich. We still move forward with courage and even abandon. We will meet any challenge head on. It is the American way. We do not plan for the adversity, the setback, the disappointment. Whatever arises will be dealt with. Certainly, it cannot be said that we plan against the contingency of destructive fire.

We can, though, with little effort.

Each of the almost eight thousand annual deaths

by fire or smoke in the United States should enrage us, for each death by fire or smoke could have been prevented. Our country has the largest per-capita fire mortality rate in the world. There are more than two million burn injuries in America each year, many of them requiring hospitalization. This should enrage us, too, for if we would only invest some time into intelligent fire-safety preplanning, and some money into simple, inexpensive fire- and smoke-alerting mechanisms, we could have the lowest rate of fire and fire mortality in the world.

The purpose of this book is to advance the cause of fire safety. This is a simple, straightforward guide book, for I believe that clever phrases and fancy technological catchwords will not save a single life in a fire—but sound, basic, commonsense information will.

In my life as a writer, I have had the opportunity to travel through much of the United States, meeting people from all walks of life, from the Hollywood movie mogul to the neighborhood banker, the television producer, the college professor and the local hotelkeeper. I use these folks as referents purposefully, for they are all generally well educated and successful professionals. Yet I found that typically these people have no real understanding of what might happen in a room, or in a hotel corridor or throughout a building full of fire. More alarmingly, none has any real sensibility of what specific action they should take if confronted by a fire in certain situations. If we spent all of our time in one kind of building, we could easily learn specifically how to act in the event of a fire in that building. However, most of our lives are spent going in and out of various kinds of buildings or environments, from wooden frame dwellings to theaters, basement discos, rooftop restaurants, hotel high-rises, garages, airports, cars, and planes. Each environment poses particular fire-safety problems, and I will illustrate these problems for you. I also aim to convince you of two things—that to be fire-conscious is to be sensible, and that sensible action in a fire or smoke situation will save your life.

A survivor of the tragic MGM Hotel fire in Las Vegas was reported to have said, "I don't think that any training would have prepared me for that fire. . . . We never heard any warning . . . or an alarm of any kind. . . ." It is understandable that a person who has lived through a horror in which a great number of people have died would say that no amount of training could have prepared him or her for such an emergency. But anyone involved in fire control or fire prevention knows that quite the opposite is true. Any reliable training or information about fire safety is better than none.

In the face of fire, the overriding consideration is to save yourself and to save those around you while n endangering others at the same time. Most of us since we were children have realized the need to remain calm in an emergency situation, for we have seen enough evidence that panic kills as surely and as quickly as flames or the swirling poison of smoke. The awful lessons of history—the Triangle Shirtwaist Factory fire, which killed 146 in 1911; the Cocoanut Grove nightclub fire, which killed 492 in 1942; the Hartford Circus Tent fire, which killed 168 in 1944; the recent Beverly Hills Supper Club fire, which killed 165 in 1977; and the MGM Grand Hotel fire, which killed 85 in 1980—are glaring and constant reminders that fire is an extraordinarily dangerous foe. Despite these tragedies, there are no serious, nationally supported fire education/fire prevention programs to help diminish the very real risk to all of us who might be helplessly caught in a fire or smoke situation.

This book can be used as a primer in fire safety education. It is designed, section by section, building environment by building environment, to increase your awareness of fire problems and to suggest ways for you to protect yourself and your loved ones from the perils of smoke and fire. The book can succeed in its fire safety mission only in direct proportion to the amount of fire consciousness it brings to your everyday, workaday life. Since there are no universal fire codes and regulations (they are enacted at the state, county, city, and even village level), I cannot recommend that certain actions be taken in all fire prevention or fire emergency situations. However, I can assure you that with reason and orderly action my commonsense approach to fire safety will help you to protect yourself and those around you in any fire emergency situation.

DENNIS SMITH'S
FIRE SAFETY BOOK

CHAPTER I
FIRST PRINCIPLES OF FIRE SAFETY

Human Reactions to Fire

How to Protect Yourself When Fire Strikes

How to Teach Your Children the Basic Rules of Fire Safety

HUMAN REACTIONS TO FIRE

What happens to human beings physically, mentally, and emotionally when they find themselves in a fire emergency?

Fire eats up oxygen, so the air in a fire is rapidly depleted of it. The air that we normally breathe contains about 21 percent oxygen, but we still can function well in levels down to 17 percent. However, once the oxygen level of air falls any lower, our functions become impaired. We are then in danger of losing muscle coordination and mental alertness, making any action more difficult to make.

Fire also produces toxic gases, among them carbon monoxide. Since these gases often have no taste or smell, we may not even be aware of them, but their effects can be devastating. For example, carbon monoxide can affect the brain so we think we are acting brilliantly when in fact we are acting in an ineffectual manner. Other gases produced by fire numb the nervous system, acting like an anesthetic and causing us to take no action at all.

Like gases, heat is another product of fire, and heat can build up rapidly. Human tolerance to heat follows this pattern:

150° F—We can tolerate this temperature for limited periods of time, depending in part on the dryness of the air.
250° F—Human tolerance is fifteen minutes.
290° F—Tolerance quickly drops to five minutes.
350° F—This temperature can be tolerated for less than a minute before the skin is irreversibly damaged.

The emotional reaction to fire varies among human

beings. Generally when fire strikes, most people are stunned for a moment or two. But then they begin to take action. Others, though, may go into emotional shock that renders them incapable of taking action promptly or sometimes at all.

Dr. Anne W. Phillips of Massachusetts General Hospital described the effects of such emotional shock on people who find themselves in a mass disaster. Of the 100 percent of the disaster population:

25 percent assess the danger correctly and take prompt action.

50 percent assess the situation properly but are irresolute. They feel considerable fear, but take no initiative. They will, however, follow a leader or obey orders.

15 to 25 percent perceive the situation imperfectly and require strong urging to respond.

A smaller group is so terrified that these people are incapable of responding to instructions.

So the real test of how you will react in an emergency situation is one I profoundly hope you will never take. Should that terrible time and event come, though, I want you to meet the situation head on with sensible and orderly action. The information about fire, smoke, and emergency situations in the following pages will help provide you with the sense and the order of fire safety.

HOW TO PROTECT YOURSELF WHEN FIRE STRIKES

1. Don't panic—concentrate instead on the steps you must take immediately.

2. Act quickly—fire is fast-acting, so every moment counts.

3. If your clothes catch fire:
   ▶ Don't run—stop where you are.
   ▶ Drop to the ground and roll back and forth.
   ▶ Keep up your arms to protect your face.
   ▶ Smother the flames.

How to protect yourself when fire strikes

4. If you burn yourself:
   ▶ Cool the burn with cold water to reduce pain and stop skin damage (not ice water—it can bring on shock). The longer you flush a burn, the better, even for as long as twenty to thirty minutes.

- ▶ Do not put butter, grease, or ointment on the burn.
- ▶ Do not attempt to clean the burn.
- ▶ Do not attempt to remove cloth that may be stuck to the burn.
- ▶ Do not break blisters.
- ▶ Cover large burns with a dry, clean towel or sheet.
- ▶ Elevate burned arms or legs.
- ▶ Call for medical help if the burn is serious.

5. If there is smoke in the air:
   - ▶ Drop to the floor and stay low.
   - ▶ Crawl to the door on your hands and knees.
   - ▶ Feel the door; if it is hot, don't open it.
   - ▶ Go to a window and try to get help from there.

What to do if there is smoke in the air

Smoke is a killer. Of every hundred fire deaths, eighty are caused by smoke inhalation, not by flames. And smoke need not be thick and rank to be lethal. In today's world of polymer plastics and woven synthetics, even a slight haze can carry chemicals or gases that can knock you unconscious in seconds. However, smoke does rise toward the ceiling, so try to stay under the smoke.

6. Never use an elevator to escape from fire. Fire moves upward, and elevator shafts invite its progress. If an elevator stalls, as it well may do in a fire, you will be trapped.

HOW TO TEACH YOUR CHILDREN THE BASIC RULES OF FIRE SAFETY

According to the National Fire Protection Association (a private, nonprofit organization based in Boston), children and fire make a tragic combination at least four thousand times a year. That is the number of children who annually lose their lives to fire. Preschoolers, age three years and under, are the most frequent fire victims. Children under five years of age make up only 7 percent of the population, yet they account for 17 percent of those killed by fire. And, of course, children in much greater numbers suffer painful and sometimes disfiguring burns as a result of fire.

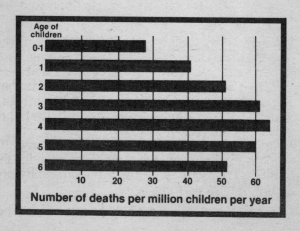

Age of children

0-1	
1	
2	
3	
4	
5	
6	

10 20 30 40 50 60

Number of deaths per million children per year

Such figures point up the desperate need to teach children both how to avoid fire and what to do if fire occurs. This fire safety training should be consistent and frequent, as children need the repetition. However, some care must be taken in how the training is done. For example, help children understand the series consequences of fire, but don't make the lessons two frightening. Children are too likely to miss the message if they grow fearful. Also, don't demonstrate dangerous fire activities, such as playing with matches. Children are too likely to imitate your behavior.

With these cautions in mind, train your children in these five basic rules of safety:

1. Teach children the dangers of playing with matches, cigarette lighters, and candles. And be sure that you keep such fire hazards out of their reach. This means any place that their inquiring minds and busy hands can take them—into purses and pockets, onto tables, and into drawers.

2. Teach them to stand well back from any open fires such as leaf fires, campfires, and barbecues.

3. Teach them the steps to take if their clothes catch fire. Have them practice stopping, dropping to the floor, and rolling over to smother the flames so that they will remember to do these things automatically. (You might call on them at unexpected moments to pretend that their clothes have caught fire and to react.)

4. Teach them to drop to the floor and crawl if a room fills with smoke. Have them practice this reaction by calling on them unexpectedly now and then to perform it.

5. Teach them to keep their hands off kitchen

ranges, space heaters, or fuel-burning stoves. Train them to understand that these are hands-off objects *at all times*, whether in operation or not. If children never touch them, they will run no risk of burns. Also, remember to keep all pot handles turned in over the stove (but not over another burner).

Keep pot handles turned in

CHAPTER II
KEEPING SAFE IN THE HOME

Go on a Fire Hazard Hunt in Your Home

Plan Your Escape

Acquaint Baby-sitters and House Guests with Fire Safety Measures

Fire Hazards and Preventive Measures

Barbecuing
Bathroom Safety
Christmas Decorations
Electric Cords
Electrical Wiring
Fabric Flammability
Fire Extinguishers
Fireplaces
Fireworks
Flammable Liquids
Furnaces
Gasoline-Powered
  Machinery

Insulation
Kitchen Safety
Matches and Lighters
Residential Sprinkler
  Systems
Smoke Detectors
Smoke Masks
Smoking in Bed
Space Heaters
Upholstered Furniture
Wastebasket Fires
Wood- and Coal-
  Burning Stoves

After a Fire Is Out

GO ON A FIRE HAZARD HUNT IN YOUR HOME

Every year 720,000 residences in the United States are struck by fire, leaving nearly 6,000 people dead and 300,000 injured. You can reduce the chances that you or your family will become part of these statistics if you take the time to do a systematic check of your home now, while there's still time to spot and defuse any potential fire hazards. Some items on the checklist might seem obvious and even clichés (such as discarding oily rags), but don't let that slow you down—all items are important.

Go on this fire hazard hunt now. And repeat it periodically.

Fire hazard check list:

☐ 1. Check your basement, attic, or other storage areas for rubbish, old newspapers, and oily rags. These are highly combustible and should be cleared out immediately.

☐ 2. Check the same areas for flammable liquids such as gasoline, kerosene, paint thinners, and alcohol-based products. (See pages 43–47 to learn how to deal with them.)

☐ 3. Check that your furnace is inspected at the start of each heating season. Clogging can cause deadly carbon monoxide gases to escape into your home.

☐ 4. Check that no old furniture or mattresses are stored near the furnace. Closness to heat can set them afire.

☐ 5. Check your water heater setting to see that

it is not above 130°F. Too high a setting can lead to tap and water scalds.

☐ 6. Check that a blowing curtain or any hanging material cannot come into contact with heating coils or gas burners on your kitchen range.

☐ 7. Check that no flammable liquids are stored near any heat sources in the kitchen.

☐ 8. Check the placement of any aerosol cans in the house to see that none is stored close to heat sources. High heat can cause aerosol cans to explode.

☐ 9. Check all electric cords for fraying. Exposed wires in appliances, lamps, and extension cords can send out fire-producing sparks as well as cause electric shock.

☐ 10. Check that no electric cords run under carpeting or across it. Room traffic frays them quickly, exposing their wires.

☐ 11. Check all electric outlets for overloading. "Bottlenecking" of electric plugs into outlets can cause overheating and start fires. (See illustration of bottlenecking on page 27.)

☐ 12. Check your light fixtures to make sure you are not using bulbs having greater wattage than called for (for example, a 100-watt bulb in a fixture calling for a 60-watt bulb). Overwattage can cause overheating, which can start a fire.

☐ 13. Check when your fireplaces were last cleaned. Chimneys should be cleaned at least once a year to prevent buildup that can lead to fire.

☐ 14. Check that you have a fire screen that covers the entire front of the fireplace to prevent flying sparks.

☐ 15. Check when any wood- or coal-burning stoves were last cleaned. Here too, a buildup can cause fire. Stovepipes should be cleaned once a month. Also make sure that the stoves are safely positioned at least three feet from the wall and have a non-combustible base beneath them.

☐ 16. Check the placement of any space heaters to see that they are well out of the way of a room's traffic patterns, draperies, and up-holstered furniture.

☐ 17. Check to see that there is plenty of air space around stereos and television sets. Too little air around them can cause overheating and fire.

☐ 18. Check all ashtrays to see that they are deep, have wide edges, and are made of metal, glass, or some other material that will not burn. A burning cigarette can easily fall backward and out of a shallow, narrow-edged ashtray.

PLAN YOUR ESCAPE

If fire should strike your home, survival depends on early detection and quick, organized escape. Early detection is especially important because more fire deaths are caused by smoke and gases than by actual flames, because smoke and gases often spread even before any flames appear. Also remember that most residence fires occur during sleeping hours. This is the time when one is less likely to be aware that smoke and flames are racing through the home.

Your best "early-warning system" lies in installing smoke detectors at strategic points in your home. These detectors are designed to set off a loud, piercing sound in the first moments of trouble, giving you the time needed for escape. (See pages 61–64 for information on choosing, installing, and maintaining a smoke detector system.)

Plan Ahead

1. See that there are two possible escapes from each part of the house. Remember that a raging fire outside a bedroom door can cut off that exist, so if the bedroom window is too high above the ground to allow safe jumping, install a rope or chain ladder. It can be tossed over the windowsill to permit quick escape. An adequate rope or chain ladder costs from fifteen dollars to about thirty dollars.

2. If you live in an apartment building, study the escape instructions provided by the building and acquaint your family with them. If no such instructions are posted, ask your local fire department to supply a plan. (If yours is a high-rise building, see pages 107–108 and 123–133 for information on surviving fire in hotels or other high-rise buildings.)

3. Work out a plan of escape with your family. See that everyone understands the need to get out (small children often think that hiding under a bed will save them) and that they know the route of escape from each part of the house. See that all escape routes always are kept clear of furniture.

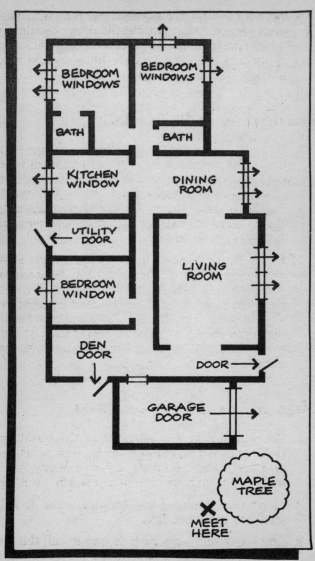

A home escape plan

4. Set up a central place where everyone is to meet after escape. Many fire deaths occur needlessly when someone, usually a parent, runs back into a burning building to look for someone who already has escaped safely.

Practice Fire Drills Regularly

1. If people, young or old, rehearse an escape plan well, they are much less likely to panic or to be confused during a fire. Run regular fire drills in which people practice staying low and crawling, feeling the doors for heat before opening them, using alternative fire exits, and meeting in the assigned place.

2. Run some of these drills after it gets dark. With children a game of fire safety can be made of sensible fire drills. They can be blindfolded to simulate darkness, but only under the strictest supervision. Parents should play, too, for in fires, the electrical systems sometimes go out, so we should all practice finding our way out in darkness.

Steps to Take if Fire Does Strike

1. Get yourself and others out of the building. Don't pause to dress or gather up your valuables and keepsakes. Neither your modesty, your money, nor your sentiment are worth your life.

2. Close doors behind you as you leave. It will slow the spread of fire.

3. Don't stay in a burning building to call the fire department. Get out and make the call from a telephone or call box *outside* the building.

4. Do not reenter the building until the fire department has said it is safe to do so.

ACQUAINT BABY-SITTERS AND HOUSE GUESTS WITH FIRE SAFETY MEASURES

Since baby-sitters are responsible for your children when you cannot be, be sure to acquaint baby-sitters with the plan of escape you have worked out with your family. Take a little extra time to show them your escape diagram, walking them through the house as you do. Help them fix in their minds where each child will be and what instructions each has received on what to do should fire strike. Acquaint them with any alarm/detection system you have, what sound it makes, and what to do should it sound. Also, show them the exact spot that you have designated as the family meeting place outside the home.

Run off several duplicates of the Baby-sitter's Fact Sheet on the opposite page. Fill in your name, address, and other pertinent information, and leave it at the phone for baby-sitters after first acquainting them with what it is and what it is for.

FIRE HAZARDS AND PREVENTIVE MEASURES

Barbecuing

Since the end of World War II, barbecuing has become a way of life in millions of American homes.

Baby-sitter's Fact Sheet

Family Name_____

Address of Home_____

Phone Number of Fire Department_____

Phone Number of Rescue Squad_____

Phone Number of Police_____

Location of Nearest Fire Alarm Box (or Neighbor's Phone)

_____

Location of Family Meeting Place

_____

Names and Ages of Children_____

_____

_____

Phone Number Where You Can Be Reached_____

The equipment ranges from tiny, single-hamburger-size hibachis to elaborate gas-heated barbecue grills equipped with electric-powered rotisseries to free-standing brick pits. The locations where these grills are used are equally varied, from fireplaces to apartment balconies to backyards and patios to picnic grounds, poolsides, and beaches.

Unfortunately, the astonishing growth of barbecuing popularity has been accompanied by an astonishing growth in barbecuing accidents. An 11-year-old Boy Scout at an annual campout had a sudden and unexpected tumble backward. He overturned a charcoal broiler and fell onto the hot coals. In just a moment, extensive second-degree burns covered his back and he had to be rushed to the hospital.

A little common sense and a lot of care can help you and your family to avoid such tragedies. Whether your equipment is simple or elaborate, there are certain precautions you should take with its use:

1. Keep children away from all barbecuing equipment.

2. Use a grill only when it is placed well away from any walls, overhanging eaves, tree branches, lawn umbrellas, and anything else the flames or rising heat waves might ignite.

3. Clean the grill frequently with water and detergent to avoid grease buildup and the dangerous grease fire it can so easily produce.

4. Use a grill only in a well-ventilated area. Burning charcoal can produce toxic fumes that can kill when they accumulate. Ventilation is particularly important if you are using charcoal in an indoor fireplace to start a fire or to cook in a hibachi. Be sure the fireplace is properly ventilated (see page 37), and keep a window opened slightly to be on the safe side. Use a larger

charcoal burner or any gas-powered grill only out of doors.

5. Avoid using fatty cuts of meat. They can lead to flare-up.

6. Never leave foods cooking unattended on a barbecue grill.

7. Keep a Class ABC fire extinguisher nearby. (See page 32 for information on fire extinguishers.)

Safety with a Charcoal Briquette Grill

1. Use only charcoal lighting fluid to start the coals. Never use any other kind of flammable liquid.

2. Douse the briquettes liberally with the starter fluid and allow them a few minutes to absorb it.

3. Stand well back when you light the coals.

4. DO NOT apply additional starter fluid to the coals once they are burning or glowing. The flames can ignite the spray and shoot back up into the container, causing it to explode in your hand. If you must rekindle the fire, use twigs.

Safety with a Gas-heated Grill

1. Read your instruction book carefully, and follow to the letter all of its instructions for assembly, hookup of gas line, cleaning, and cooking.

2. Do not allow the electric cord connected to the rotisserie to contact the broiler, as such contact can burn the cord and expose wires.

3. Do not allow the gas line or hose to contact the broiler, as such contact can burn a hole and cause a dangerous gas leak.

4. Turn off the gas tank valve when the barbecue is not in use and store and the tank only in a cool, dry area—never inside the barbecue or in direct sunlight.

Bathroom Safety

Fire hazards in the bathroom center around electric appliances, space heaters, hot water and hot tap scalds, and aerosol cans. Each year twenty-six hundred people, mostly children and old people, are injured, either by scalding water or by contact with excessively hot taps. Of this number, an average of forty die as a result.

Here's what you can do to ensure that no accident in your family is included in one of these statistics:

1. Use great care with any electrical appliances in the bathroom. Avoid touching anything electrical like a hair dryer if your hands are wet, if you are standing in water, or if you are standing at a sink in which there is water.

Keep electric space heaters out of the bathroom

2. Keep electrical appliances out of reach from the bathtub.

3. Never use an electric space heater in the bathroom. Such closeness to water simply is inviting disaster.

4. Do not place aerosol cans near hot water pipes or other heat sources in the bathroom.

5. Keep your water heater setting lower than 130° F. This not only lowers the chance of scalds but also saves energy.

6. Don't allow small children to be alone in the bathroom, where they might come into contact with hot water or hot water taps, and check the temperature of their bath water with your elbow before putting them into the water.

Keep children away from hot water taps

Christmas Decorations

At Christmas, play and make good cheer,
For Christmas comes but once a year.

This was the hearty advice that appeared in *The Farmer's Daily Diet* in 1557, a piece of advice that millions of Americans are eager to follow each year. Yet for thousands of them, it ends in needless tragedy. In the United States alone, careless actions at Christmastime send thirty-five hundred individuals to hospital emergency rooms as a result of fire-related accidents.

To make sure that you and your family enjoy nothing but "play" and "good cheer" this coming and every Christmas, keep these cautions in mind.

Choosing a Christmas Tree

If you are choosing a natural tree, dryness is the biggest hazard to fire safety. The fresher the tree, the less likely it is to be dry when you bring it into the home. If you live near a tree farm that allows you to cut your own tree, you can be quite certain of the tree's freshness. But if you buy a cut tree, you must exercise greater care. Don't be misled by a high green color—the tree may have been sprayed. Instead, take these steps:

1. Tap the tree lightly on the ground. Do many needles fall off? If they do, the tree is too dry. Don't buy it.
2. Bend a few needles with your finger. If they break easily, it is also too dry. Pass it by.

If you are choosing a plastic tree, make sure that it bears a seal of approval from a nationally recognized testing laboratory and that it is made of fire-resistant material. This does not assure that it will not burn, but at least it will not catch fire easily.

If you are choosing a metal tree, you do not risk a fire hazard. But you must exercise care in how you light it. Use only floodlights. Don't attempt to place lights on the tree itself. If you do, you risk electrically charging the tree and causing severe electric shock. Remember, though, that floodlights can become extremely hot, so keep everyone, especially children, away from them.

Caring for a Christmas Tree

1. When you bring a natural tree home, keep its base in water until you are ready to set it up.

2. Cut the butt end of the tree diagonally one or two inches above its original cut.

3. Place it in a firm holder with a wide base and fill the holder with water to cover the cut line. Refill the holder as necessary to keep the water up to this level, checking it every day.

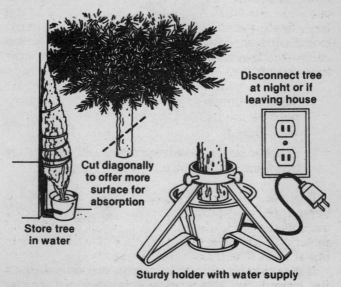

Disconnect tree at night or if leaving house

Cut diagonally to offer more surface for absorption

Store tree in water

Sturdy holder with water supply

Christmas tree care

4. Keep the tree only until it retains most of its needles. If needles are falling off in great quantities, get rid of it because it's dangerously dry.

5. Place a Christmas tree, either natural or plastic, safely away from any heat source. Also, when placing a Christmas tree, remember never to block an exit.

6. Christmas trees can be dangerous to small children and animals if they fall, so keep the little ones away from the tree to prevent unnecessary accidents.

Using Christmas Lights

1. When you buy lights, whether for indoor or for outdoor use, see that they bear the seal of approval from a nationally recognized testing laboratory guaranteeing their safety.

2. Check your Christmas lights, indoor and out, each year before using them. Are any of the cords frayed, exposing wires? Are there loose connections? Are there broken sockets? If there are, either throw out the light strands or repair them thoroughly.

3. Never use indoor lights for outdoor lighting. They simply are not designed for it. Outdoor lights must be weatherproofed and clearly labeled for outdoor use.

4. Fasten all lights securely to the tree, and don't let the bulbs contact the needles or branches directly. Don't let the lights come into contact with any other flammable materials either, such as curtains or decorations.

5. See that all Christmas tree decorations, such as ornaments, tinsel, and decorative rope, are labeled as noncombustible or flame-retardant.

6. Don't overload Christmas light strands onto an extension cord. Use no more than three strands per cord, and don't let a cord stretch across any area where it might be tripped over. Keep all electric wires away from the tree's water supply.

7. When you go out of your home or retire for the night, do not leave Christmas tree lights on. Unplug them from the wall outlet.

8. Never, never, no matter how much it may remind you of childhood or carry on European tradition, light wax candles on or even near a Christmas tree—it simply is too dangerous.

Electric Cords

Have you ever been through a blackout due to an electrical power outage? If you have, it probably was the first time you became aware of how truly dependent you are on electricity.

Picture your home for a few moments. How many fixtures or appliances do you have that are plugged into your electrical system? A typical listing might include the following:

Lamps and light fixtures
Radio
Television
Stereo
Air conditioner
Electric space heater
Refrigerator
Kitchen range (if electric)
Barbecue with electrically powered turning spit
Toaster/toaster oven
Kitchen mixer and/or blender
Iron
Coffee maker
Electric frying pan
Electric fondue pot
Electric clocks
Hair dryer
Electric tools
Recreational gadgets such as electric train transformers

All of these items have one thing in common in addition to running on electricity—they all have electric cords. Some of the cords are hidden away, such as those for a refrigerator or range. But some are out in plain view and subject to wear and fraying. So are any extension cords you might use in connection with them. These "out in the open" cords present two potential and grave dangers—electric shock and fire. (Electric coffee makers, skillets, and fondue pots, when overturned by an accidental tug at their cords, present two more dangers—scalds and grease burns.)

The most frequent victims of accidents involving electric cords are children between the ages of six months and four years. After all, they are often down on the floor, where they can get at electric cords easily. Left unattended, children are likely to suck and chew on electric cords. They also are likely to tug at the cords, sometimes exposing wires. And the moisture from their diapers and from water they spill can bring on severe electric shock and burns.

To prevent tragedies caused by electric cords from striking you or anyone in your family, take these precautions:

1. Inspect all electric cords regularly for signs of wear. Has fraying exposed any wires? Have the cords worked their way loose from their plugs? If so, replace the cords immediately or have them replaced professionally. And be sure that if you do any replacing yourself, you use a cord that is clearly labeled to show that it is specifically designed for the type of appliance in question.

2. If you use any extension cord, see that it matches the electric rating of the appliance for which you intend it. An extension cord for an air conditioner is not suitable for a lamp, and vice versa.

3. Never overload a wall socket, and do not "bottleneck" electric cords into an extension cord socket. Overloading a cord socket with too many appliances can ignite the cord's insulation.

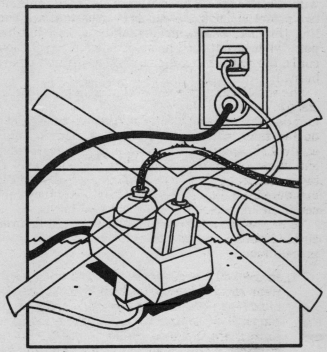

Don't overload sockets and inspect cords for signs of wear

4. If you have any fixture or appliance plugged into an extension cord, see that it fits snugly, with no part of the prongs exposed. The same is true for plugs fitting into wall sockets.

5. Don't allow unused wall sockets or unused sockets on extension cords to be uncovered—

either use electric tape to cover them or fit them with safety caps. (Children can electrocute themselves by sticking an object such as a nail file into an open socket.) Also, if you have an appliance or fixture plugged into an extension cord permanently, tape the plug to the socket to prevent children from unplugging it. Remember, the exposed extension cord socket has the same potential as an exposed wall socket.

6. Do not run any electric cords under rugs or carpets, under any door, or over any nails. All these contribute to wear on the cord.

7. Do not run any electric cords near radiators or other high heating source.

8. Keep the electric cords of kitchen or bathroom appliances away from any water source. If electric cords must be near water, see that they are waterproof.

9. Never let appliance cords hang freely over counters where small children will pull on them.

10. Never wrap an electric cord around an appliance—(for example, an iron) until the appliance is cool to the touch.

11. See that any electric cord you use outdoors— for example, for electric barbecue spits or outdoor Christmas lights—is designed specifically for outdoors. The cord must be covered with the proper insulation to make it waterproof.

If you have an electrical fire:
▶ Turn off the current immediately.
▶ Use water only after the current is off; otherwise you may electrocute yourself.

▶ Use a Class C, BC, or ABC fire extinguisher, if handy. (See page 33.)
▶ Get out if the fire grows and call the fire department.

Electrical Wiring

An estimated 4.3 percent of all residential fires are caused by fixed but faulty electrical wiring. Obviously, you cannot see into your walls or ceilings to spot electrical wiring flaws, but fortunately there are telltale signs that something is wrong. If you see any of the following signs of faulty wiring in your home, call in an electrician for immediate inspection and repair:

◊ Dimming or flickering of lights

◊ Regular blowing of fuses or regular tripping of circuit breakers

◊ Overheated electrical outlets or cover plates on electrical outlets

◊ Electrical sparks or glowing from electrical outlets

Such signs as these may stem from a number of causes. Perhaps electrical wiring has deteriorated or perhaps you are using fuses or circuits that are too much for the wiring to handle safely.

Your wiring may not be faulty by itself, but you may be calling on it to do too much, with disastrous results. Never use a bulb with more wattage than a light fixture can handle. If a fixture says, "Use only 60-watt bulbs in this fixture," don't use 100-watt bulbs in it. Such overwattage can cause overheating, and overheating can cause fire.

Fabric Flammability

Look around the room in which you find yourself right now. How many fabrics do you find close by? First of all, there are the clothes you are wearing. Loose-fitting clothing is more susceptible to catching fire than tight-fitting clothing. Then, too, perhaps you are sitting on a piece of furniture that is upholstered, either with a natural cloth or a synthetic product such as vinyl or Leatherette. Perhaps draperies or curtains hang at the windows and there are rugs or carpeting on the floor. If you are in a bedroom, there are bedspreads, blankets, and mattress covers near you. All of these fabrics have the potential to burn, quickly in some cases, more slowly in others.

The dangers of fabric flammability are at least two-fold. First, there is the danger that if they ignite, they can burn your body as well as spread fire. Second, there is the danger that their burning can produce toxic gases that can knock you out, making your escape impossible. Here are a few of the literally thousands of cases involving fabric flammability that injure and kill each year:

◊ A six-year-old finds matches and begins playing with them. He drops a lighted match on his shirt. It catches fire and 40 percent of his body is covered with burns.

◊ Mr. G. spills some gasoline on his clothes as he is filling the tank of his lawn mower. A spill of gasoline falls on the concrete floor of his garage. He had neglected to turn the motor off, and a spark ignites the spill. Mr. G. bends to smother the flames, and his clothing erupts into fire. He survives, but only after a long hospitalization.

◊ Mrs. J., wearing a large-sleeved kimono, reaches for the salt and pepper shakers, which have been negligently left at the back of the stove. A sleeve brushes by a lighted burner and erupts into flame, and Mrs. J. becomes critically injured.

◊ The decor in a large, popular nightclub includes a synthetic covering on the stairs and banquettes. As a sudden flash fire moves through the club, the burning synthetic covering produces nitrous oxide. It knocks out many of the patrons even before they can move to excape. They die of carbon monoxide poisoning.

In the past few decades, the federal government has established standards to reduce the dangers of fabric flammability, mainly in carpets, rugs, mattresses, and children's sleepwear. These government standards do not assure that such fabrics are flameproof but that the fabrics manufactured within the standards resist flames better than ordinary fabrics. So when shopping for any of these fabrics, check to see that they are labeled as having met standards for flame resistance. And when laundering such products, such as children's sleepwear, follow the laundering instructions printed on them so as not to lose their flame-resistant properties. When shopping for any upholstered furniture, learn what you can about their qualities in regard to fire. How flame-resistant might they be? What gases might be created if they catch fire?

When choosing clothing, it is good to keep in mind that 50 percent of all burns are caused by clothing fire. Therefore it is useful to bear in mind these facts:

1. Tightly woven fabrics burn more slowly than loosely woven ones.

2. Heavier fabrics burn more slowly than lighter ones.

3. Smooth fabrics burn more slowly than fabrics with a fluffy pile or a brushed nap.

4. Close-fitting clothing is less likely to catch fire than loose-fitting clothing. Bell sleeves, ruffles, long nightgowns, and loose shirt tails are particularly susceptible to contact with ignition sources such as space heaters, range burners, and fireplaces.

5. Nylon, acrylic, and polyester fabrics all do a fairly good job of resisting ignition, but once they do catch fire, they melt, which can cause serious burns.

Fire Extinguishers

Do-it-yourself fire fighting can be a dangerous enterprise, but quick and sensible action often can keep a minor fire from developing into a major one. Having fire extinguishers handy around your home can put you in a position to take quick and sensible action. But which kind should you buy? How many? Where should you place them for maximum effectiveness? How can you make the best use of them in an emergency?

Choosing a Fire Extinguisher

Home fire extinguishers generally are labeled to show the kinds of fire for which they are designed. Look at the illustration on the following page to see what those labels mean:

Class A—This designation stands for fires involving materials such as wood, paper, cloth, and certain plastics.

Class B—This shows fires involving grease or flammable liquids.

Class C—Electrical fires come under this heading.

Extinguisher classes

The extinguishers themselves come in four classifications and contain the proper materials to fight the different classes of fires.

Class A—These extinguishers contain water to be expelled by a carbon dioxide gas cartridge or air pressure. Or they contain a soda or acid solution.

Class AB—These contain foam to fight Class A or Class B fires but not Class C fires.

Class BC—These contain dry chemicals that are effective against Class B and Class C fires, but not against Class A fires.

Class ABC—These extinguishers also use dry chemicals to fight fire, but they work against Class A fires as well as against Class B and Class C fires.

Because fires of different sorts may possibly occur in any one room (for example, paper fires, grease fires, or electrical fires are possible in a kitchen;

cloth fires and electrical fires are possible in a bedroom or living room), probably your best bet would
be to acquire the all-purpose Class ABC extinguisher.

Weight is another important consideration in
choosing a fire extinguisher. Not surprisingly, the
heavier the extinguisher the greater its fire-fighting
capacity, because it contains more extinguishing material. A ten-pound extinguisher can extinguish
twice as much fire as a five-pound extinguisher. But
you do want to take into account the people who are
liable to be using it and be sure to choose a weight
that they can manage in an emergency.

The power of a fire extinguisher is designated by a
number paired with its class rating. The designation
1A means that the extinguisher can put out a test fire
of 50 pieces of 20-inch-long wood 2 x 2's. A rating of
2A shows that the extinguisher can put out a fire
twice that size. A rating of 1A:10BC (a common rating for a 5-pound, 16-inch-high extinguisher) shows
that it can put out a 1A fire or a fire 10 times as strong
as 3¼ gallons of naphtha burning in a 2½-square-foot
pan. The usual running time for home fire extinguishers runs from 8 to 25 seconds.

See that the model or models you choose have a
pressure dial so that every month or two you can
check to see that the pressure is constant and no leak
has occurred. Also see that any extinguisher you
choose is easy to use. Its safety mechanism must
unlock easily, and its activating mechanism, like a
depressor or a squeezable handle, must not require
undue strength. Finally, choose a fire extinguisher
that comes equipped with a hanging device so that it
is always in the same place and can be reached with
ease.

Placing Fire Extinguishers in Your Home

Home fire extinguishers are not expensive (considering their potential for preventing major losses),

costing from about twenty dollars to fifty dollars apiece, so you may want to place several around the house. The kitchen is perhaps the most logical place. So is any room in which you have a fireplace, space heater, or heating stove. A bedroom is a likely place, too, because of the vulnerability of a sleeper to fire. If you store flammable liquids in the basement or cellar or the garage, these areas are logical for a handy fire extinguisher.

Wherever you decide to place your extinguishers, do not put them too close to the place where a fire might occur. For example, in the kitchen, don't put an extinguisher right next to the range. If a fire occurs there, you do not want to have to reach through the flames to get the extinguisher. Instead, place it near the door. You can quickly reach it there, and if you see that the fire is too dangerous, you can escape.

Using a Fire Extinguisher

1. With grease fires, stand ten to twelve feet back from the flames. If you stand closer, the pressure of the extinguisher's stream is likely to send the grease flying, spreading the fire.

2. With other fires, stand as close as you can without endangering yourself.

3. Aim the nozzle at the base of the fire, activate the extinguisher, and sweep the nozzle across the fire base.

4. Keep a door behind you. Don't ever let a fire get between you and your means of escape.

5. Don't try to fight too big a fire. If a fire continues to spread, get out! Call the fire department. (If someone else is around, get that person to call the fire department the moment you spot the fire, as you begin to use the extinguisher.)

Using a fire extinguisher

6. Familiarize the members of your family with the use of fire extinguishers. It might well be worth the cost of a recharge—five dollars to seven dollars—to do a "dry run" out of doors with them, to give everyone the feel of a fire extinguisher in action.

7. Take a fire extinguisher to a dealer for recharging as soon as it has been used (even if you've used it for only a few moments) or if you think it has a leak. Otherwise, it will not contain enough pressure when you need it.

8. Take a fire extinguisher into an extinguisher service shop at least once every six years for a thorough checkup. It pays to maintain your investment in an extinguisher.

Fireplaces

You are a king by your own fireside,
as much as any monarch on his throne.

These words, written nearly four hundred years ago by Cervantes in his preface to *Don Quixote*, capture the spirit of warmth and security that fireplaces conjure up for many people. A fireplace brings to mind visions of peace, quiet contemplation, and perhaps even a bit of romance. On a less fanciful level, though, in these days of high energy costs, a fireplace also brings to mind its original, utilitarian function— to produce heat.

Both for reasons of sentiment and of utility, fireplaces are making a real comeback in home renovation and construction these days. But with their resurgence has come a growing list of problems. Each year more than fifteen thousand people are treated in hospital emergency rooms for injuries stemming from fireplace accidents. And each year, the number of residential fires that spring from fireplace misuse grows.

If you have or are thinking of having a fireplace in your home, you should understand its workings. Look at the cross section of a fireplace on the following page. Note especially the fireplace proper, for it is here, of course, that the actual burning takes place. Above it is the *damper*, through which the smoke and gases pass upward into the *flue*. This action takes place through a process called *convection*. As the logs heat up, convection draws fresh air from the room and into the logs, creating a draft. This draft pulls the smoke and gases from the fire up into the flue and out the *chimney*. The hotter the fire grows, the more fresh air it draws in and the more gases it sends up the flue, creating a still stronger draft.

As this process continues, a roaring fire results, but it must not be allowed to roar out of control. The damper, a metal plate controlled by a handle, keeps the fire at the proper level. Partially closing the damper reduces the draft and thus keeps the fire at a safe level. Of course, the damper cannot be closed completely or the fire will be starved of the oxygen it

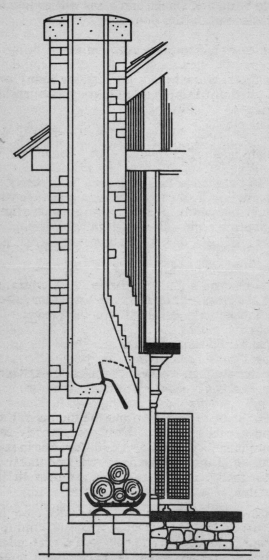

Fireplace and chimney cross section

needs to burn, and smoke and gases will go into the room instead of up the flue.

Major Causes of Fireplace Accidents

◊ Clothing such as a loose-fitting sleeve or a skirt brushes too close to the flames and ignites.

◊ Sparks fly out from a fire burning too hot or burning damp or resinous wood and ignite rugs or furniture in the room.

◊ A flammable liquid is used to kindle or rekindle a fire. Or flammable liquids are used in the vicinity of a fireplace in which a fire is burning and the vapors ignite.

◊ Inadequate venting causes deadly carbon monoxide to seep into the room.

◊ Defective chimneys, the result of either improper construction or deterioration, allow flames to spread to surrounding areas.

Avoiding Fireplace Accidents

First, be sure your fireplace is a fireplace and not a realistic-looking fake.

1. Have your fireplace inspected annually. Have an expert make sure that it is properly constructed for use as a fireplace, that it is in good repair, that it is properly vented, and that there is no dangerous creosote buildup in the chimney that can catch fire.

2. Have the chimney cleaned annually, more frequently if you use it often. Creosote buildup is highly flammable and may well catch fire long before you are aware of it, causing a house-

threatening chimney fire. Such buildup also can block a chimney, preventing toxic fumes from escaping and sending them instead into the room.

3. Use chimney guards so that no animals or birds can nest in the chimney, blocking it.

4. Never use flammable liquids on or anywhere near a fireplace, and don't store any of them in its vicinity.

5. Keep the area around the fireplace free of any flammables, and don't place furniture or rugs close by the hearth.

6. See that the damper is open before you start a fire, and use the damper properly to keep the fire under control.

7. Follow the safety precautions described on the packaging if you use artificial logs. These logs are made of compressed sawdust and wax, and they burn much hotter than natural logs. Use only one at a time, and never use them together with natural logs. The two simply do not mix!

8. See that a room is properly vented (open a window a little) if you are going to burn charcoal or coal in the fireplace; both give off potentially toxic gases.

9. Don't burn rubbish, especially polystyrene packaging, in a fireplace.

10. Use a fire screen that completely covers the opening around the fireplace. It will keep dangerous sparks from flying out.

11. Teach children to keep well back from a fireplace and alert all family members to the danger of clothing fires near a fireplace.

12. Don't leave a fire unattended.

13. Be sure a fire has gone out completely before you leave it to go to bed.

14. Dispose of ashes only when they have thoroughly cooled; they can stay "live" for as long as twenty-four hours, making them potential fire starters. And use only a metal container for them.

15. Keep an extinguisher suitable for wood fires (Class A) handy to the fireplace.

Fireworks

Nothing more can be said about setting off fireworks except DON'T! There is probably no American adult who doesn't know personally or hasn't heard of someone who has suffered a personal injury from fireworks. Each year eight thousand victims enter hospital emergency rooms for treatment for burns to the eyes, blown-off fingers, or other physical calamities. Conservative estimates place the death toll from fireworks accidents at ten per year.

As a result, the federal government has prohibited the sale of some hazardous fireworks—cherry bombs, aerial bombs, M-80 salutes, and firecrackers containing more than two grains of powder. (To give you an idea of the wallop that fireworks can pack, a new model called the Super M-80 has the explosive power of an eighth of a stick of dynamite.) And many states have banned or limited the sale of so-called Class C fireworks, including fountains, Roman candles, rockets of many varieties, mines and shells, smaller firecrackers, and sparklers. But even in states that have banned fireworks, people do get and use them.

If your state allows fireworks and you insist on using them, at least follow these sensible precautions:

1. Never allow younger children to play with fireworks. They simply cannot understand the danger or, in many cases, the steps to take when emergency strikes. Children are truly the unwitting victims of fireworks. Even sparklers cannot be considered "safe" for younger children. Sparklers burn at high temperatures and can set fire to clothes.

2. Carefully supervise the use you allow older children to make of firecrackers. Be sure that they and you understand and follow the directions on fireworks containers. And be sure that the children follow the directions without "horsing around" and approaching others with fireworks.

3. Use fireworks only outdoors and in a clear area well away from buildings or flammable materials.

4. See that other people are well out of range when you light fireworks.

5. Keep a bucket of water nearby whenever fireworks are in use. Use it to douse any malfunctioning fireworks or in any other emergency.

6. Do not handle or try to relight fireworks that have malfunctioned. Soak them thoroughly, and then throw them away.

7. Never light fireworks inside a container, especially one made of glass. Flying particles can have the force of bullets shot from a gun.

8. Store fireworks only in a cool, dry place that is locked and away from children.

It is thought that fireworks originally were invented by the ancient Chinese to frighten away ghosts. Today, fireworks carelessly handled should frighten all of us away.

Flammable Liquids

Think about your home for a minute. Do you have any of the following lying around it? Gasoline? Kerosene? Lighter fluids? Oil-based paint? Alcohol-based paint thinners? Volatile substance-based products such as nail polish remover, airplane glue, or rubber cement? Charcoal starter fluid? If you do, you have potential fire starters in your midst, ones whose action you should understand. The U.S. Consumer Product Safety Commission divides these liquids into three categories:

Extremely Flammable Liquids—Gasoline, white gasoline (used in camping stoves and lanterns), and certain adhesives and wood stains fall into this category. Each of these liquids can produce ignitable vapors at room temperature and even well below it (under 20° F). These vapors are heavier than air, and that means they can flow along the floor or down a flight of stairs. If they come into contact with an ignition source, such as

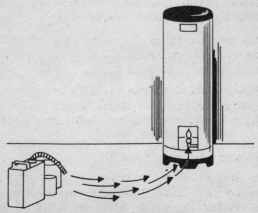

Extremely flammable liquids can produce ignitable vapors

a water heater pilot light or a burning cigarette, they can flash into fire with explosive force and rush the flames back to the liquid itself.

Flammable Liquids—Paint thinners and certain paints fall into this category. They too produce ignitable vapors, but they require higher temperatures to do so.

Combustible Liquids—Kerosene, diesel oil, fuel oil, furniture polishes, and oil-based paint are all included here. The label *combustible* on a container tells you that such products will burn once they are ignited, but they are much less likely to simply catch fire than the materials in the two categories above.

Of course, all of these products have their uses around the home, and that is why we keep them. But they must be used and stored properly if you are to limit them to their positive uses and guard against negative results. And the results can be not only negative, but tragic as well. Of the Americans who suffer burns from flammable-liquid accidents each year, fully 70 percent must undergo painful and extensive treatment in the hospital.

Gasoline is the most dangerous flammable liquid of all. It is made to explode! What would you guess the explosive power of one gallon of gasoline is? That of one stick of dynamite? Two sticks? Ten sticks? A gallon of gasoline has the explosive power of *thirty* sticks of dynamite! Think about what that would mean to your home or car if you have a gallon of gasoline stored in either place.

Defusing the Dangers of Gasoline

1. Never store gasoline in your house. If you must keep it around, for use with lawn or garden machinery, store it in a well-ventilated garage or

shed away from your living quarters. Good ventilation is an absolute necessity to prevent vapor buildup.

2. Use a proper storage container for gasoline. This means that a glass container is definitely out—think of the potential danger of a broken bottle. So is a plastic container or a metal container with a spout. Use only a metal safety can designed specifically for the purpose. It should have a flame arrester and a pressure-release valve that lets out excess vapor. (The buildup of such vapor can cause a can to leak or to burst in a heated place.)

3. Never store gasoline in the trunk of a car. (One Boston man learned this lesson in a tragic way. His wife and three small children were stopped in their car at a toll booth when a truck plowed into the back of the car. A gallon can of gasoline stored in the trunk exploded, killing the man's wife and the three children.)

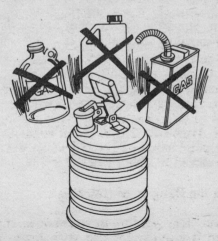

Proper gasoline storage containers

4. Never use gasoline around an ignition source, even a distant one—remember the traveling power of gas vapors.

5. Use gasoline only in a well-ventilated setting so that dangerous vapors cannot build up.

6. Never use gasoline as a starter for charcoal; use only a starter fluid intended for this use.

7. Keep gasoline locked away from children.

8. Do not use gasoline as a cleaning fluid for clothing, metal parts, or anything else. Friction can cause gasoline to burst into flames.

9. Do not refuel power mowers or any other gas-powered machinery when they are turned on or when any of their parts still are warm. First turn off the machinery and wait until it is cool to the touch. (More information on safety with gas-powered machinery begins on page 48.)

Using Other Flammable Liquids Safely

1. Use flammable liquids only in well-ventilated areas, to prevent dangerous vapor buildup.

2. Do not use flammable liquids around any ignition source, for the same reasons you would not use gasoline near one. And do not smoke when you are filling a cigarette lighter with fluid.

3. Store flammable liquids only in tightly capped metal containers that are clearly and accurately labeled, and keep them out of the reach of children.

4. Do not store flammable liquids near any heat sources, such as hot-water pipes, kitchen ranges, furnaces, or space heaters.

5. Rags that have been saturated with flammable

liquids should be treated just like the liquids themselves. Keep them in tightly closed metal containers.

Furnaces

Each year more than eight thousand people enter hospital emergency rooms for treatment of injuries associated with central heating and floor furnace accidents. Essentially, these accidents follow one of three patterns:

◊ A child crawling on the floor encounters a hot floor grate, resulting in severe burns and inability to move away. Or an adult trips and falls, landing on a hot surface.

◊ A man attempts to light a pilot light that has gone out in his furnace. As he strikes the match, he ignites the vapors that have accumulated since the light went out. His clothes catch fire and he has burns over 50 percent of his body.

◊ A woman stuffs a towel into the intake vent of her furnace in order to cut off the flow of cold air coming from it. Lacking the necessary oxygen to burn its fuel properly, the furnace produces large quantities of deadly carbon monoxide, which kill the woman.

All of these accidents could have been prevented. To prevent them in your home, follow these simple rules:

1. Cover any floor grates with screens so that there is no possibility of a child or an adult coming into contact with a burning hot surface.

2. Select a furnace with a "safety pilot" if you are buying a new furnace. It will automatically shut off the flow of fuel if the pilot light goes out.

3. Don't attempt to relight a pilot light yourself unless you are sure it is a "safety pilot." Instead, call the service or gas company to do it for you.

4. Do not store flammable liquids near the furnace. The pilot light can ignite their vapors, if any are seeping out.

5. Do not store old mattresses or furniture near the furnace—the closeness to heat can set them afire.

6. Have your furnace and its flues checked once a year, before the heating season begins, to see that air intake and exhaust are working properly and that there is nothing blocking the flues that could cause deadly carbon monoxide fumes from them to enter the house.

7. Take care if you are insulating your house that the insulation does not cover any air intakes or exhausts—such interference also can send deadly fumes into your house.

Gasoline-Powered Machinery

The American mania for labor-saving devices has replaced the old rotary-blade, muscle-powered lawn mower and the arm-powered ripsaw with fast-acting, gasoline-powered machines. The whirring of these machines can be heard in any residential neighborhood on any weekend in the spring, summer, or fall. In colder climates, gasoline-powered snow plows and snow blowers have replaced shovels. Such power tools have cut down the labor demanded for home upkeep, but they also have produced everyday fire hazards that did not exist a few short decades ago.

Obviously, the fire hazards these machines present involve their fuel—gasoline. The dangers of gasoline have been pointed out elsewhere in this book, under "Flammable Liquids" (pages 44–46). One gallon of gasoline has the explosive power of thirty sticks of dynamite. To prevent a possible burn injury, or even worse, follow these steps whenever you are fueling gasoline-powered machinery:

1. Don't smoke when you are refueling.

2. Don't refuel when there is any ignition source nearby. Gasoline vapors can be ignited even by a distant source.

3. Never refuel indoors—in a garage or a basement, for example. Refueling is strictly an outdoor operation, which should be performed on bare ground.

4. Don't fill the tank to the very top. Leave a little room for fuel expansion.

5. Wipe up any gasoline that spills.

6. Move at least ten feet away from the spot at which you do the refueling before you turn on the machine. This takes you away from the gasoline vapors that have arisen during refueling.

7. Never run any gasoline-powered machines indoors. They can produce carbon monoxide fumes that can be lethal in a confined area.

8. If you must refill the tank in the midst of a job, wait until the machine cools before refueling.

9. Never leave a piece of gasoline-powered machinery unattended. And keep children and pets away from it.

10. Don't let any part of your body come into contact with any area of the machine that becomes hot during operation. You could get a painful burn.

11. Let the machine cool before storing it. If, when operating a machine, you must leave it for a short period, be sure to close the fuel shut-off valve.

12. Review the directions for storing gasoline properly that appear on pages 44–45 under "Flammable Liquids."

Insulation

As heating bills climb ever upward, many people turn to insulation as a heat-saving and thus money-saving measure. Then too, the government has encouraged home insulation by providing tax incentives for those who install it. If you are thinking of taking this route, good for you! But there are a few things you should know about fire safety and insulation.

First, it's a good idea to have your electrical system inspected by an electrician before installing insulation, especially if your home shows any of the signs of faulty wiring listed on page 29.

Second, choose a type of insulation that is naturally fire-resistant, such as fibrous glass, or a kind that has been treated with fire-retardant chemicals. Be sure that whatever insulating material you choose meets the building codes in your area. If you choose cellulose insulation, be sure it has a label that shows it meets the U.S. Consumer Product Safety Commission's safety standards.

Whether you hire a contractor to do the job or elect to do it yourself, be sure that these cautions are closely followed, to assure fire safety:

1. Don't smoke while installing insulation.

2. Keep insulation at least three inches away from the sides of any recessed light fixtures and from the sides of any junction boxes of surface-mounted lights. You may want to place a barrier around such fixtures or boxes to assure that proper clearance is maintained. (Running insulation over light fixtures or junction boxes entraps heat and thus can bring on fire.)

Fixture and box separated from installation

3. Recheck the wattage called for on any light fixtures near insulation, and see that the bulbs in use do not exceed that wattage. Overwattage is

another cause of overheating that can lead to fire.

4. Keep insulation away from furnace exhaust flues, water heaters, space heaters, or any other heat-producing devices.

Kitchen Safety

Remember that the kitchen is the work area of the home, and good safety sensibility is as important here as in any industrial workshop. Keep things neat and orderly, and always be aware of the dangers that the presence of children brings.

It is impossible to give an accurate estimate of the number of fires that begin in American kitchens each year because so many go unreported. It is known that cooking accidents cause about 130,000 fires that are reported to fire departments each year. But probably from 5 to 10 times this number go unreported. Of the fatal fires in American residences each year, fully 16 percent begin in the kitchen.

Here are just a few of the ways kitchen accidents occur:

◊ Sixty-year-old Mrs. B. was wearing a robe with loose-fitting sleeves as she reached to stir a pot on the back burner of her kitchen range. Her sleeve contacted the fire from the front burner and ignited her robe. By the time she could smother the flames, her arm was completely covered with burns.

◊ An electric coffee maker sat on the kitchen table, percolating merrily. Two-year-old Davy, playing on the floor, reached for the cord that hung down over the table edge and pulled. He toppled the scalding coffee over himself and suffered burns over one third of his body.

◊ Jessie was using paint thinner to assist her in repainting a chair in her kitchen. As she worked, vapors accumulated to the point that the pilot light on her range ignited them and flashed the flames back to the thinner. The kitchen soon was engulfed in flames, fatally trapping Jessie and soon spreading to the rest of the house.

Avoiding Kitchen Accidents

These tragedies and many others could have been avoided if a dozen simple rules were obeyed:

1. Don't wear loose-fitting clothing near the kitchen range or broiler burners. It is sensible, too, to tie back long hair.

2. Don't use a small pan over a large burner. The exposure of hot coils or flames increases the danger of clothes contacting fire.

3. Avoid storing anything over or on the kitchen range. This helps to prevent you from reaching up over lighted burners and igniting clothes.

4. Don't use dish towels or paper towels as pot holders—they can easily contact fire and burst into flames.

5. Don't allow spilled grease to collect around burners or in the broiler pan. Grease buildup is almost sure to ignite when next you turn on the fire.

6. Don't leave a pan with grease in it unattended over a fire, and don't let it heat to smoking. Grease can burst into flames in a moment.

7. Hot liquids burn more people than fire does. Carry a pot or a cup of hot liquid with great care, and see that no children are around when you do. Also, make sure you use a teapot that will protect you against a scald from rising steam. It should not have a removable lid, it should be fillable from the spout, and it should be large enough to keep the spout a safe distance from your hand.

8. Always turn pot handles inward so they do not stick out from the range. Protruding pot handles invite overturning.

9. Don't let electric cords hang down within reach of children.

10. Don't place hot liquids near the edge of a counter or table. Children can too easily reach up and pull them down.

11. Don't store flammable liquids in the kitchen, and don't use them there unless the pilot light is turned off and the room is well ventilated. Remember, never store drain acids, lye, bleaches, solvents, chemicals, and poisons under the sink, where they are accessible to children.

12. Don't use your range or oven burners to heat the room. This often leads to fire.

If a fire should occur, take these steps.
For a grease fire on the stove:

1. Turn off the burner.

2. Cover the pan with a lid to smother the flames.

3. Don't pour water on a grease fire. It may only splash the grease and spread the flames; the same is true of using baking soda on it.

Turn off gas (or heat)   Smother with lid   Never use water

Extinguishing a grease fire

4. Don't try to pick up a flaming pan and carry it to the sink to run water on it. You are likely to burn yourself and you risk spreading the flames to other parts of the kitchen.

5. If you use a fire extinguisher on the flames, stand back ten to twelve feet. Standing closer will create pressure that can send the flames shooting out.

For a grease fire in the oven or broiler:

1. Close the oven or broiler door.

2. Turn off the heat.

Matches and Lighters

A young mother, busy in the kitchen, happened to glance into the living room where she knew her 2½-year-old son was playing. She screamed in horror, for not only was her son playing with a book of matches, but also the couch behind him was completely ablaze. She scooped up the boy and raced out of the apartment, screaming, "Fire! Fire!" to the neighbors in the seven-story building.

Neighbors fled their apartments as thick smoke made its way down a hallway. In the confusion and, no doubt, in an attempt to help, one of them grabbed the boy from her arms. "Where is my boy?" she cried out in anguish.

An eighty-four-year-old neighbor, hearing her cries, assumed that the boy must be trapped in the now blazing apartment. "He opened the door and the second he did," reported an approaching fire fighter, "flames blew out with a whoosh right into his face." Flames were licking at the old man's body as the fire fighters reached him.

Though the would-be rescuer was critically burned, he did survive. Another elderly neighbor was not so lucky. He died after collapsing in his struggle to get out.

Children playing with matches is an old and often repeated story in tragic fires such as this one. It is a story that is retold thousands of times each year. But adults too can be careless with matches, and old people sometimes can be clumsy with them. Matches, so common in everyday life, must be recognized for the ever-present and powerful ignition sources they are, and they must be treated with all the care that their potential danger demands.

Using Matches with Safety

1. Do not use matches that are damaged or discolored. They probably have been exposed to dampness and their heads are likely to break apart when struck, throwing off lighted particles that can strike flesh or clothing.

2. Close the matchbook cover before striking. This still is good advice even now that striking strips are placed on the back of matchpacks.

3. Close the matchbox when using wooden safety matches. A spark from the struck match can fly into an open matchbox and set the entire package ablaze.

4. Don't carry more than one matchbook in the same pocket or purse at any one time.

5. Hold a match an arm's length from the body when striking it to avoid contact with flying sparks. And strike with a motion going *away from*, not toward, the body. Never, never strike a match close to your face, even if it is windy.

6. Don't discard a match until the flame is out and the match is cool to the touch.

7. Don't strike a match unless the action has your full attention. Accidents often happen when people are distracted; you may miss the fact that a burning particle has fallen on something that it might ignite.

8. Never strike a match in the vicinity of flammable liquids.

9. Use care in storing matches. Don't leave them in damp places. Don't leave matches in pockets from one season to the next, as they get old and dangerous that way. And don't carry matches in a suitcase when traveling. Their rubbing together can ignite clothes.

10. Always bear in mind that young children and matches must never mix. Keep matches out of the reach of children. Avoid brightly colored matchbooks, a real lure to childish eyes. And teach older children to follow the rules for using matches that appear above.

Using Lighters with Safety

Like matches, lighters are a wonderful convenience. And like matches, they are potential hazards, whether they are fueled with butane or lighter fluid. By following these rules, you can cut down the danger:

1. Strike a lighter well away from your face. This way you will avoid accidents should a flame larger than you expect flare up.

2. Concentrate on what you are doing when you strike a lighter. An unwatched flame can lead to a serious situation.

3. Check a lighter for leaks frequently. A leaky lighter in your hand invites burned fingers, and a leaky lighter in your purse or pocket adds to the flammability of your clothes.

4. Check the flame level of lighters often. The adjustment mechanism can be hit accidentally (in one's pocket, or during housecleaning) and can be reset for a higher flame.

5. Use great care when refilling a lighter with fluid. Pour the fluid only out-of-doors or in a well-ventilated room. DON'T SMOKE at the same time. And wipe up any spill that might occur, either on the lighter itself or on another surface.

6. Take an extra moment to make sure that the fire is out before you put a lighter away. Carelessness here can result in a burned hand or ignition to clothes.

7. Don't use a lighter around flammable liquids.

8. Teach children the dangers of lighters. Lighters are every bit as dangerous as matches and produce the same accidents. Keep table lighters out of the reach of young children.

Residential Sprinkler Systems

A relatively new innovation in home fire protection is the residential sprinkler system. For years, schools, office buildings, warehouses, and factories have incorporated sprinklers as part of their fire safety systems, saving lives and property as well as lowering fire insurance rates. (See a fuller treatment of the operation and reliability of sprinkler systems on pages 105–107.) Now residential sprinkler systems are being developed to help cope with the fires that break out in one of every ten American homes each year.

Residential sprinkler systems have the overriding advantage of limiting home fires to the room of origin. They are designed to go off in one fifth the time that commercial and industrial systems take, which of course means that they can stop a home fire dead in its tracks.

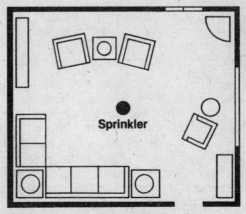

Sprinkler

Living room sprinkler site

Nevertheless, many people continue to object to residential sprinkler systems, largely through lack of knowledge about them. How knowledgeable are you about the subject? To find out, take the "I.Q. test" below, which was put together by the Federal Emergency Management Agency.

Test Your Home Sprinkler System I.Q.

Here are five statements about home sprinkler systems. Are they true or false?

1. When one sprinkler goes off, all the sprinklers activate.
   False! Only the sprinkler over the fire will activate. The sprinkler heads react to temperatures in each room individually. Thus, fire in the bedroom will activate only the sprinkler in that room.

2. A sprinkler might very probably go off accidentally, causing severe water damage to a home.
   False! Records that have been compiled for well over fifty years prove that the likelihood of this occurring is very remote. Furthermore, home sprinklers are specifically designed and rigorously tested to minimize the probability of such accidents.

3. Water damage from a sprinkler system will be more extensive than fire damage.
   False! The sprinkler system will severely limit a fire's growth. Therefore, damage from a home sprinkler system will be much less severe than the smoke and fire damage if the fire had gone on unabated or even the water damage caused by water from fire-fighting hoses.

4. Home sprinkler systems are expensive.
   False! Current estimates suggest that, when a home is under construction, a home sprinkler system could cost about 1 percent of the total

building price. Residential sprinkler systems could use standard piping and hardware with domestic plumbing.

5. Residential sprinklers are ugly.
False! The traditional, industrial-type sprinklers as well as sprinklers for home use now are being designed to fit in with almost any decor.

Well, how did you do? Does a sprinkler system now sound a little more feasible for your home?

Smoke Detectors

In the early hours of an April morning, Peter M. awoke to find that a short circuit in his electric blanket had triggered a fire. At first he tried to extinguish the flames, but soon they were out of control. Only then did he try to awaken other members of his family. Five of them were able to break through a first-floor window and escape. But for Peter M.'s son and grandson asleep upstairs, it was too late. By the time they awakened, the smoke and heat were just too intense, and they were totally overcome by them.

Had their home been equipped with smoke detectors, they probably would not have died. At the first sign of smoke, the detectors would have gone off with an ear-splitting noise, giving father and son the precious moments they needed.

It is estimated that as many as two thirds of the people who die in home fires would have survived if they had had a smoke detection system. Yet only 10 percent of American homes are equipped with smoke detectors, even though they are relatively cheap and easy to install. Smoke detectors come in a price range of about fifteen dollars to fifty-five dollars.

Types of smoke detectors

Heat detectors also are readily available to warn of fire. However, since smoke generally is detectable before heat, smoke detectors are faster-acting than heat detectors.

Choosing a Smoke Detector

Obviously, the most important thing to look for in a smoke detector is that it works. So be sure that any detector you buy bears a seal of approval from one of the nationally recognized testing laboratories.

Next, there is a choice which involves the two types available on the current market. Either type should provide timely warning, but it is good for the consumer to know the differences between them. The first is the ion chamber detector. It detects both visible and invisible products of combustion by using a radioactive source to trigger a chain of events that sets off the alarm. The second type is the photoelectric detector. It detects only visible products of combustion because it relies on a light-sensitive cell reflecting the smoke particles to set off the alarm.

You also will have to decide between a battery-powered detector (which uses a single nine-volt battery) and one that operates off your home's electrical system. The advantages of the battery-powered detector are that it is easier to install, it does not need to be placed near an electric plug, and it will continue to function in the event of a power blackout. The advantage of the electric system detector is that there is no risk of batteries going dead unnoticed. To capitalize on the advantages of each type and to minimize the disadvantages, you might want to use a combination of the two in your home.

Finally, you may want to choose a model that allows for interconnection with other detectors in the house. This means that as soon as one detector sets off an alarm, all the others will join in.

Installing a Smoke Detector

1. Place detectors on or near ceiling areas where smoke is likely to rise—for example, at the top of a stairway.

2. Place detectors near where people sleep. Remember, you are most vulnerable to fire at night.

3. Place at least one detector on each level of a multilevel house.

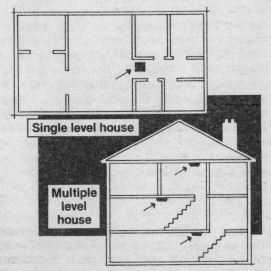

Proper smoke detector locations

4. Don't place a smoke detector in the kitchen, where cooking grease and smoke are liable to set off a "false alarm." And don't place it near an air-conditioning or heating vent, which might blow the smoke away from the detector, keeping it from doing its job.

5. Place detectors on ceilings (at least six inches away from where the ceiling joins the wall) or on inside walls (at least six to twelve inches below the ceiling). Such placement takes advantage of the principle that smoke rises, and it keeps detectors high enough to be away from accidental bumps and curious children.

6. For extra protection, especially if your family sleeps with bedroom doors closed, you may want to place a detector in each bedroom.

Maintaining a Smoke Detector

1. Test a smoke detector at least once every month—or better still, once a week. For an ion chamber detector, hold a lighted candle six inches below it and let the flame burn. For a photoelectric detector, put out the candle but let its smoke drift up from six inches below. If the alarm goes off within twenty seconds, the detector is working. Fan away the smoke to quiet the alarm. (Many detectors come equipped with a test button, but to be on the safe side, do the candle test anyway.)

2. Replace bulbs for a photoelectric detector as soon as they burn out.

3. Replace batteries at least once a year, whether the warning device for weak batteries has sounded or not.

Smoke Masks

Smoke can be a killer. Coroners estimate that eighty out of a hundred fire deaths are caused not by flames but by smoke inhalation. Smoke is a deadly substance made up of gases that in a fire emergency situation can impair vision, cause great irritation to the lungs, cloud thinking, and numb the nervous system. Next to panic, smoke is your greatest enemy when trying to protect yourself in an environment that is burning.

Recently, a new product has been introduced on the popular consumption market—the emergency smoke mask. This device features a built-in filter/respirator that can extend safe breathing time in a smoke-filled atmosphere by three to four times. The device itself is a noncombustible hood that slips over the head. A rubber strap secures it around the chin, and a fog-resistant window permits the wearer to see out. It is lightweight—only ten ounces—and it fits into a small carrying case. It also is available in a wall mount, from which it can be taken easily. At present, the emergency smoke mask is priced at about thirty-five dollars, but that figure may well come down as production escalates.

Emergency smoke mask

At this early stage in the history of emergency smoke masks, it is perhaps prudent to offer a few cautions about them. First of all, using a smoke mask should not make you overconfident about what you can reasonably do in the event of a fire. Don't let wearing a smoke mask tempt you to enter a smoke-filled situation when you might be safer staying where you are. Second, wearing a smoke mask is no guarantee against oxygen depletion. If fire causes the oxygen level of the air to drop below 17 percent, you will be no more impervious to the lack of oxygen with a mask than without one. These smoke masks carry no oxygen or air supply with them, unlike the self-contained breathing masks that fire fighters wear. I would suggest, though, that they are better than no smoke protection at all, and so I've included an emergency smoke mask in the survival kit you should have with you when traveling (see pages 134–135). It also is a good idea to keep one of these in your bedroom at home.

Smoking in Bed

DON'T SMOKE IN BED! How many public service television spots do you suppose you have seen that delivered that message? Probably as many as you have seen warning that if you drink, don't drive. Yet each year millions of people in the United States ignore the warning. And the results are staggering.

◊ Smoking is the second largest cause of death in residential fires (the first is woodburning stoves), and nearly a third of the deaths are related to smoking in bed.

◊ There is a high correlation, especially in the sixteen-to-sixty-year-old age range, between the use of alcohol and fire deaths, and a large proportion of these deaths result from smoking in bed.

If you figure that this couldn't happen to you because a fire in your bed would wake you in time, think again. The smoke from your smoldering mattress can knock you out long before the heat of the fire might rouse you. Smoke from a smoldering mattress has knocked out thousands before you.

You might ask all your friends and loved ones who smoke: Is that last cigarette of the day, smoked in bed, worth the risk that it might be your last cigarette ever?

Don't smoke in bed

Space Heaters—Electric, Gas, Oil, and Kerosene

The recent emphasis on saving energy has convinced many householders that space heaters offer a way to heat a room efficiently and economically or to add additional warmth to a poorly heated room.

Space heaters can do these jobs, if they are used properly. Used improperly, space heaters can have effects such as these:

◊   Clothing coming into contact with space heaters can burst into flames.

◊   Overturned heaters can set fire to rugs or floors.

◊   Brushing up against a space heater can cause severe burns.

◊   Blowing window curtains coming into contact with a space heater can catch fire.

The occurrence of accidents such as these has been growing in recent years, as the number of space heaters in use has grown. Each year, at least fifty-eight hundred people require emergency treatment in hospitals for burns related to space heaters. You can cut down risk of such accidents in your home by following these recommendations:

1. Keep heaters away from curtains, drapes, and upholstered furniture. All these are easy targets for fire.

2. Keep heaters out of the lanes of traffic flow through rooms. One false step can overturn a heater or cause a painful leg burn.

3. Keep a window open an inch or more when using a fuel-powered space heater. The burning action can create carbon monoxide, which is deadly if fresh air is not also coming into the room.

4. Train children to stay away from space heaters. See that they learn never to touch them, in operation or not.

5. See that any electric space heater has a grille

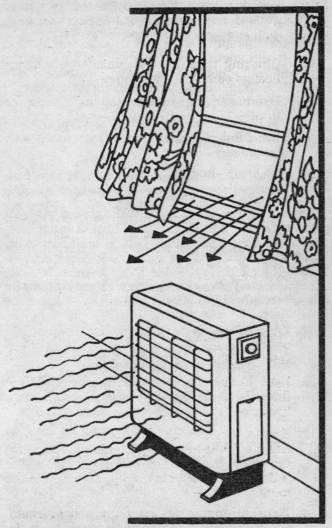

Open a window when using fuel-powered space heaters

or other protection around the coil, to prevent contact with flammable materials or with skin.

6. See that any electric space heater you buy has an automatic tip-over switch that shuts off the current when it is knocked over—this helps to prevent ignition from occurring if the heater is tipped over accidentally.

7. Never use an electric heater in a bathroom or near a sink. Electricity and water are a deadly combination.

8. Make sure that the venting system of a fuel-powered space heater is in proper working order; have a professional check it if you are not certain. Loose joints, undersized or clogged vents, and cracks encourage the accumulation of deadly carbon monoxide in the room.

9. Use only the fuel for which the space heater was designed. Don't use oil in a kerosene heater or vice versa.

10. Never use flammable liquids in the same room with a space heater. A space heater is an ignition source for flammable vapors.

Upholstered Furniture

Just a few years ago, a well-known entertainer bid his guests good-bye as they left his Los Angeles apartment after a reportedly long and fairly liquid evening. The next the guests learned of their host was that his apartment had been gutted and the charred remains of a man had been found on the couch in the living room. After long hours of medical examination and checking of dental records, identification was finally made. The remains were those of the celebrity, Jack Cassidy.

Fire officials pieced together what might have happened. Cassidy had lain down on the couch and lit a cigarette. He dozed off and the cigarette ignited the couch's upholstery. Unnoticed, the fire smoldered, producing deadly smoke that the sleeping man inhaled. From then on he never moved, and the flames gathered force and swept through his apartment.

Cigarettes are the most common ignition source of fires in upholstered furniture, but nonsmokers are not immune. Several other heat sources also are causes—space heaters, heating stoves, wall and floor furnaces, and sparks from the fireplace. Once these sources come into contact with upholstered furniture, they have the same effect as a lighted cigarette—fire that can smolder for hours before it finally bursts into flame, or fire that can ignite much more quickly.

Upholstered furniture offers potential dangers beyond just flame and the normal smoke of incomplete combustion, as if these were not deadly enough. The polyurethane foam or other synthetic material with which it is stuffed or covered can produce toxic gas when ignited. These too can kill, and they can do it at a distance. There are numerous cases on record in which fire was quietly smoldering unnoticed in a piece of upholstered furniture in a living room or den when the family went off to their bedrooms. Toxic gases then seeped into the air and spread, killing or injuring the unknowing family members in their sleep.

Preventive Steps

1. Keep upholstered furniture well away from all heat sources.

2. If you have a fire in the fireplace, be sure to have a screen across it so that no sparks can leap out into the room, noticed or unnoticed.

3. Check any electric cords that run near uphol-

stered furniture to see if they are frayed and therefore capable of throwing fire-starting electrical sparks.

4. See that your ashtrays are deep and have wide sides, to lessen the chance that a cigarette can fall out of one of them onto upholstered furniture.

5. Check ashtrays before you go to bed at night to see that no cigarettes are left burning in them.

6. Check the cushions of the chairs in which anyone was smoking, before you go to bed, to make sure no cigarettes are left smoldering in upholstered furniture.

Check the cushions of the chairs in which anyone was smoking before you go to bed

7. Be extra careful with cigarettes if you have been drinking alcohol or taking medication. Either one is liable to make you drowsy, especially in the comfort of an upholstered chair or couch. And never smoke lying down.

If a cigarette does fall onto a piece of upholstered furniture:

1. Pick up the cigarette immediately.

2. Extinguish any sparks or glowing ashes.

3. Take off any cushions and examine them and the areas they covered to see that no sparks have spread.

4. Check below the furniture to see that no sparks have fallen there.

If you find that a fire has been smoldering in a piece of upholstered furniture:

1. Get out.

2. Call the fire department.

Remember that a smoldering fire in furniture or in a mattress can flare up at you like a blowtorch if you move it in such a way as to increase the amount of oxygen it is using. So be careful around a smoldering fire. Never bend over it for any reason.

Wastebasket Fires

Fire occurs all too frequently when a smoker carelessly throws a still-lighted match or a still-smoldering cigarette butt into a wastebasket. Or an overly neat housekeeper empties a messy ashtry into a wastebasket before all the butts are cold. (This is one case in which neatness definitely does not count.) In the enclosed space, amid crumpled papers, the fire source can get flames going in seconds.

To avoid such accidents, use good common sense to see that no cigarette or match leaves your hand or your ashtray until it has had time to cool completely.

But if a wastebasket fire should occur, follow these rules:

1. Act quickly.

2. Do not take time to run for water to put out the fire.

3. Do not attempt to carry the flaming wastebasket out of the room. You run too great a risk of spreading the flames as well as burning yourself.

4. Instead, smother the flames with a blanket or pillow or coat.

Smothering a wastebasket fire

Wood- and Coal-Burning Stoves

It was Christmas Eve as Anita and John F. arrived at their country home to enjoy the holidays. John threw some logs into the woodburning stove in the living room, set them afire, and went out to the kitchen to fix some Christmas cheer. When Anita came into the living room, she was shocked to see that the stovepipe was glowing red with heat. Outside, a shower of sparks cascaded from the chimney.

Creosote from the wood had built up in the stovepipe and had started a fire there, sending sparks up the chimney. Luckily, a snow covering kept the sparks from igniting the roof, and the fire department was able to put out the pipe fire quickly and safely.

Not everyone is as fortunate as Anita and John F. Each year, fires like this one, involving wood- or coal-burning stoves or free-standing fireplaces, send fifteen hundred people to the hospital in the United States. Indeed, in the Northeast especially, such heating devices are a growing cause of fire deaths.

Avoiding Stove Hazards

1. If you are putting in a stove, have it installed by a qualified person who is familiar with building inspection and fire department requirements for stoves. Be sure that the stove you select has been approved by a nationally recognized testing organization and that the stovepipes have been designed specifically for it.

2. Be sure that the stove is vented outside through a masonry chimney or one that is approved as an "all fuel" chimney; that it is placed on a brick platform, fireproof stoveboard, or other noncombustible material; and that it is at least three feet from any wall. Use of these steps prevents overheating and ignition of floor or walls. Also see that the connectors running through the wall or ceiling are properly protected so that they do not overheat and ignite surrounding surfaces.

3. Have the stove inspected once a year to see that linings are intact; that the stove is adjusted and clean; that the chimney is free of

creosote, soot, or other blockage; and that there are no faulty parts or cracks. Such cracks can allow deadly carbon monoxide to escape into the room. If you use your stove frequently, stovepipes should be cleaned once a month.

4. See that the stove is placed away from draperies and furniture and out of room traffic lanes. Direct contact with a heated stove can cause ignition of materials and burns to people.

5. Keep kindling and newspapers away from the stove to avoid combustion.

6. Keep a window open slightly when using the stove to provide enough oxygen for proper combustion and to prevent carbon monoxide poisoning.

7. Use only the fuel for which the stove was intended. Don't use coal in a woodburning stove, or vice versa. Don't ever use a flammable liquid to start or "stoke" a fire, as this can bring on an explosion. And don't use charcoal—it can produce deadly carbon monoxide.

8. Don't store any flammable liquid in the same room with a stove. The heat can ignite flammable vapors and flash a fire.

9. Keep a stove fire at moderate heat, neither too hot nor too cool. A fire that is too cool produces flammable gases that can explode when the stove door is opened, and it also builds up highly combustible creosote in the chimney.

10. Use only wood that is properly dry. Green wood builds up flammable deposits as well as corrosive deposits that can damage the metal in the stove and flue.

11. Don't use a stove to dispose of trash, especially paper and polystyrene.

12. Transfer ashes from the stove only when they are cool. Hot ashes may stay "live" for more than twenty-four hours, making them potential sources of ignition; and use only a metal container, never a cardboard box.

13. Don't ever hang clothes near a stove to dry, as they can easily catch fire.

14. Keep a fire extinguisher suitable for wood or coal fires (Class AB or ABC) handy to the stove.

If a fire does grow too hot in a stove, causing the stovepipe to glow red, take these steps quickly:

1. Close the stove dampers completely.

2. Partially close the stovepipe damper.

3. Put a few shovelsful of cool ashes on the fire.

4. Or use the fire extinguisher according to its directions.

5. Call the fire department.

Remember that the proliferation of stove sales and use in the United States has caused a proliferation of fires and fire deaths. To install or maintain a stove improperly is almost certain to end in tragedy.

AFTER A FIRE IS OUT

If the worst does happen and you do have a fire in your home, you will face the often heartbreaking and even traumatic task of salvage, cleanup, and beginning again once the fire is safely out. Wipe your tears, square your shoulders, and follow the steps below that apply to your situation.

Securing the Fire Scene

1. Do not try to reenter the premises until the fire department gives permission. Fire can easily rekindle, so wait until the fire fighters assure you that reentry is safe. They will have made certain that there is no fire hidden in walls or elsewhere.

2. Call your insurance agent to report the situation. If you are the homeowner, you may well need immediate help in boarding up broken windows and holes made to vent the flames. If you are a tenant, you will need to report your losses.

3. If you are in need of temporary housing, contact your local disaster relief services—the American Red Cross or the Salvation Army.

4. If you must leave the fire site, take your valuables. And before you leave, try to see that any necessary boarding up has been done, to help protect your property from vandalism and from bad weather. Also contact the local police to ask them to keep an eye on your property.

Getting Back to Normal

1. The fire department will have checked to see that your utilities—gas, water, electricity—are either safe or are disconnected before they leave. To be on the safe side, have an electrician check your household wiring before you have

electricity turned back on. The wiring may have been damaged by water. Before you have any other utilities reconnected, have a service representative check to see that the apparatus is in proper working order. Don't attempt to reconnect utilities yourself.

2. Have an expert check for structural damage. The fire may have weakened floors or roofs.

3. Discard any food, beverages, or medicines that have been exposed to heat, soot, or smoke, and do not open any cans that bulge, are dented, or are rusty. Food from the freezer that is still solidly frozen can be salvaged, but don't try to refreeze anything that has begun to thaw.

4. If you have a safe, don't attempt to open it until several hours have passed. A safe can hold intense heat for a long time, and if it is opened before it has cooled sufficiently, the heat can combine with entering air to cause the contents to burst into flames.

5. Keep receipts for everything you buy and for all the work you have done. You will need them for insurance purposes. It is also helpful to have photos of your jewelry and other valuable items. A safe-deposit box in the local bank can be used to protect these records.

Cleaning Up the Mess

1. To get rid of the smell of smoke in washable clothing, place them in a solution of two tablespoonsful of sodium hypochlorite or four to six tablespoonsful of trisodium phosphate mixed with one cup of chlorine bleach and one gallon of water. (Be sure you test any colored clothing before putting it in such a solution.) Rinse with clear water and dry.

2. Use this same solution to wash walls, furniture, and floors, wearing rubber gloves as you work. In washing walls, work from the floor up to lessen chances of streaking. Rinse with clear water. Do not attempt to repaint until thoroughly dry.

3. Put mattresses out in the sun to dry and to air out.

4. Wood furniture and fixtures require especially careful treatment following water damage. Get expert advice on dealing with them.

Try not to grow too disheartened. As you work through the dreadful "before," keep the bright "after" picture in your mind, and concentrate on how much more pleasant "normal" will seem to you when you get back to it again.

CHAPTER III
KEEPING SAFE IN PUBLIC PLACES

Hospitals and Nursing Homes

Restaurants, Nightclubs, and Discos

Schools

Stores and Shopping Malls

Theaters and Movie Houses

HOSPITALS AND NURSING HOMES

Health care is a growing business in the United States. Each year approximately 1 American in 6.7 goes into the hospital for some form of treatment. And as the geriatric population (those of age seventy-two and over) grows, the number of nursing homes grows with it, to handle those elderly patients who do not have someone to care for them at home and who no longer can care fully for themselves.

The vulnerability to fire that hospital and nursing home patients have is clear. Hospital patients may well be weak or, worse, immobilized. Nursing home patients may well be elderly, highly nervous, easily confused, or unable to take action when emergency strikes.

Fortunately, fire safety standards for hospitals today are very high. The last major hospital fire, taking a toll of seventy-four, occurred back in 1949 at St. Anthony's Hospital in Effingham, Illinois. Unfortunately, fire safety standards for nursing homes often are not so high, and multiple-death fires in them average at least three a year.

What can be done to reduce this number and thus reduce the danger both to yourself and to your loved ones? Look at the two fire emergencies in nursing homes described below to find some answers.

A Study in Contrasts

The scene is a one-story plus attic nursing home in a remote part of Ohio. The time is November 23, 1963. The nursing home, a reconverted toy factory constructed mainly of concrete block walls with some combustible parts, contains eighty-four elderly patients. Eight months earlier, it had passed inspection by state authorities, who had found that it complied with existing state laws. However, it has no fire detection system, no local manual alarm, no sprinklers, no emergency lighting system, and no water supply to fight fire. The only equipment it has are three fire extinguishers. There are no written emergency procedures with which nursing home personnel have been acquainted. The nearest fire department is 7½ miles away.

Shortly before five in the morning, a rolling steam table used to serve breakfast causes a short circuit in the electrical system, which is overloaded and which was improperly installed. As the flames spread, an attendant tries to call the fire department, but the telephone wires, which run overhead where the flames are, now are out of commission. Four passing motorists race in, tell the attendant to get the patients out, and start using the extinguishers to fight the fire. One motorist runs out and stops another motorist to tell him to call the fire department from a neighboring house.

By this time, the flames have raced throughout the undivided attic area. The building fills with heavy smoke, and the lights go out. The attendants and the motorists begin taking patients outside. But soon the entire roof is ablaze and it becomes impossible to reenter the building. Of the eighty-four patients, sixty-three never make it out alive.

The scene is a ranch-style, one-story nursing home in Las Vegas, Nevada. The time is July 20, 1977. Two hundred patients are having their breakfast, either in a breakfast room or in their own rooms. As a repairman works on an electrical line he causes a short circuit that brings on a flash fire. The sprinkler head located nearest to the fire activates itself immediately. A maintenance man working nearby pulls the manual alarm. A staff member, a veteran of the fire drills run every month, calls the fire department, which also has been alerted by a fire alarm system tied to it from the nursing home. The alarm has activated the smoke partition doors, closing them immediately. The well-drilled staff members calmly and efficiently follow the evacuation procedure and get the patients outside.

The smoke grows thick and black as an emergency generator automatically turns on all exit and overhead lights, which helps visibility a little. Staff members continue evacuating the patients, covering their faces with bedclothes to cut down smoke inhalation. As the staff evacuates each room, they close its door behind to retard the spread of fire and smoke.

Within minutes, the nursing home is completely emptied and the fire is out, extinguished by the single sprinkler head. There is no loss of life, and property damage is estimated at a thousand dollars.

The lessons learned from this study in contrasts could not be clearer. Fire safety in hospitals or nursing homes is built on three foundations:

1. Fire-safe construction

2. Fire safety measures

3. Fire-trained personnel

If you must go into the hospital or arrange for a nursing home occupancy, check into the fire safety record and the fire safety provisions of the health care facility you are considering. (The local fire department can help in securing this information.) Remember, you or the person you are placing probably will be somewhat impaired in protecting your or his or her own safety in a fire emergency, so choose a place that looks out for fire safety for the patient. As you consider a hospital or nursing home, use the following checklist:

☐ 1. Does the construction meet rigid fire safety codes? Is the building soundly constructed for exactly the purpose it is supposed to fulfill?

☐ 2. Does it have each of the following:
   a. a heat and smoke detection system?
   b. a fire alarm system that is near at hand and that immediately alerts other people on the scene as well as the fire department?
   c. a sprinkler system throughout?
   d. automatically activated smoke partition doors?

☐ 3. Is trash picked up and disposed of frequently and carefully? (Trash collection areas have been found to be the most frequent place for fires to begin in hospitals studied.)

☐ 4. Is soiled laundry taken away immediately (rather than left in the patient's room) and placed in closed chutes or containers?

Is soiled laundry removed immediately?

☐ 5. Are hospital clothing and furnishings chemically treated so that they are flame-retardant?

☐ 6. Are stairwells and passageways kept clear of accumulated laundry or other hospital materials? (This is important both for unobstructed evacuation and for a lessening of combustible materials lying about.)

☐ 7. Has an emergency evacuation plan been specifically drawn up, and have the health care facility's personnel been thoroughly and regularly drilled in it? (It is impossible to stress the importance of this point. According to a 1979 study, "Almost every [health care facility] fire that has involved injury or death can be traced back to failure of personnel to follow established [emergency] plans."

Here are a few safety pointers for health care facility patients:

1. Keep the number of combustible materials in your room to a minimum. All those cards and knickknacks may be cute and cheery, but they also may be highly combustible.

2. Once again, though not for the last time, be careful with smoking materials. (Recent information has shown that of the hospital fires studied, 29 percent were caused by smoking, far and away the greatest single cause.) Do not smoke when you are taking medication. It can make you drowsy and careless. And don't smoke in bed.

3. Have a nurse acquaint you with the fire evacuation procedure you will be expected to follow.

4. Keep your door closed, especially when you are sleeping. It will retard the spread of fire or smoke into your room from another location.

RESTAURANTS, NIGHTCLUBS, AND DISCOS

November 22, 1919—A second-floor dance hall in Ville Platte, Louisiana; a couple of hundred merrymakers are celebrating the weekend in the frame structure. Downstairs a fire begins in a combination grocery store-restaurant and spreads quickly to the second floor. As the crowd rushes to escape down the single staircase, the stairs collapse under the unexpected weight. A total of 25 die in the fiery crush.

April 23, 1940—The Rhythm Club in Natchez, Mississippi; 700 or more people are crowded into a long, narrow dance hall, a single-story structure—120 feet long and 38 feet wide—made of corrugated iron sheets. At 11:15 P.M., a fire breaks out at the hamburger stand at the front of the dance hall. It leaps to the decorative gray Spanish moss hanging from the ceiling joists and begins to burn rapidly. The burning moss drops down, igniting people's clothes and creating a hedge of fire that blocks patrons from reaching the dance hall's single exit at the front of the building. They surge instead toward the rear of the hall, to try to escape the flames.

Along the side and rear walls there are 18 windows, but they have all been boarded up to discourage people from entering the hall without paying. Escape by means of them therefore is impossible. For those patrons who have been able to reach the vicinity of the exit, there also are barriers to escape. This way out is partially obstructed, and doors leading to it open inward instead of outward. People soon pile against them, making their opening impossible. By this time, the heat is intense; the noncombustible iron walls and ceiling trap the superheated gases within the building.

The fire department responds to the alarm within minutes and extinguishes the fire quickly. But 207 patrons are dead, most of them piled up against the rear wall, the victims of suffocation or of trampling.

November 28, 1942—The Cocoanut Grove nightclub in Boston, Massachusetts, a reconverted garage divided by plywood walls into a series of rooms; 1,000 people are jammed into the basement and ground-floor area built to accommodate 600 at most. At 10:00 P.M. a small fire begins in a faulty light fixture and quickly spreads through the decor in the basement cocktail lounge.

In a few moments, 200 people are trying to exit the lounge by way of a narrow stairway leading up to the main floor. The lights go out, and fire flashes over the heads of those trying to escape, cutting off the stairway. Some people have not even tried to leave their tables. They already have been overcome by the toxic fumes produced by burning Leatherette. As patrons try to exit the ground floor, they find one door welded shut. Two other doors open inward, and a wall of people pressed up against them makes the doors impossible to open. The only other exit, a revolving door, soon is jammed by a fallen victim's leg. The death toll—492, nearly half of those present.

January 29, 1956—The Arundel Park, Maryland, amusement hall: 1,100 people are in the hall. Outside there is some innocent cooking going on, under an overhanging eave. The eave catches fire and soon the hall is ablaze. Because it is a ground-floor structure with several well-marked exits, most people are able to escape with their lives, although hundreds are burned. Still, 11 victims die.

February 7, 1967—Dale's Penthouse Restaurant atop the 11-story Walter Bragg Smith Apartment Building in Montgomery, Alabama; 50 customers are in the restaurant along with several restaurant employees. Shortly after 10:00 P.M. a fire begins in an unattended coatroom. Employees try to fight the fire themselves with a fire extinguisher, but the flames soon race through partitions, made partly of plywood, into the bar and then into the dining room and kitchen. The only means of exit from the penthouse are two self-service elevators and a stairwell behind them, reached through a corridor.

No one seems to have warned the diners to leave, although two groups, becoming alarmed, exit by means of the elevator. By this time the hostess tries to lead a group of customers to the stairwell, but smoke and heat have cut off the corridor. She takes them to the kitchen, where they all perish. Other customers break windows to get out on a small balcony, where they hope they can be rescued. But the smoke and the heat are too great. In the aftermath, fire fighters find 25 dead.

June 30, 1974—Gulliver's, a disco in Port Chester, New York, part of a shopping center on the Connecti-cut-New York border; 500 young adults are enjoying themselves in the dimly lit rooms that are separated by plywood partitions. Next door, in a children's playroom connected to a bowling alley, a disturbed young man sets a fire. The flames travel to the disco, where the fire is discovered, and at 1:00 A.M. the bandleader approaches the microphone to request that everyone leave in an orderly manner. An orderly evacuation begins, but soon thick smoke is rolling through the disco and the lights go out. Those pa-trons who are on the dance floor, which is about 5 feet lower than the other room, have to climb up a 5½-foot-wide stairway to get out. A bottleneck quickly occurs there, and several people lose consciousness before making it up the stairs.

When the fire department arrives, fire fighters put on self-contained breathing apparatus and enter the smoke-filled, intensely hot building. They succeed in bringing out 30 young people safely, but after the fire is extinguished, they find 24 dead. All the victims died from smoke inhalation.

December 18, 1975—The Blue Angel nightclub in New York City. At 1:30 A.M. only 50 patrons remain of the 200 who had been in the nightclub earlier. They begin to smell smoke, but it is not until 2:00 A.M. that employees find the source—a fire in the stage area that they try to put out themselves with water and milk. The fire begins to spread to the stage curtains, but the fire department still has not been called. The patrons finally start to leave, stopping for their coats along the way. The smoke thickens and the electrical power fails, causing great confusion and severely hampering evacuation.

It is not until 2:12 A.M. that an alarm finally is turned in to the fire department. This delay costs 7 people their lives. The official report on the fire concluded, "The deaths were the result of failure to abide by the cardinal rules of fire safety: 'Call the Fire Department immediately' and 'Evacuate the premises.'"

October 24, 1976—In the Bronx, New York; 55 members are enjoying an evening of drinking and dancing in the second-floor premises of a private social club that have neither been designed nor licensed for such purposes. The single fire escape is blocked by a locked metal gate. Several of the windows have been barred, bricked up, or covered. There is a single stairway leading to the first floor.

In the early hours of Sunday morning, a would-be suitor grows angry because his girlfriend is enjoying herself with others at the social club. He and two of his friends go to the club and throw gasoline on the only stairway and ignite it. A large ball of fire rolls up the stairs and bursts into the room. The patrons on the second floor quickly discover that neither the stairway nor the fire escape is available to them. So in their fright, they line up before the remaining unblocked windows to jump, but the flames and smoke race through before many of them can reach the windows. They flee back into the rest rooms or behind the bar. Firefighters find 25 bodies in these locations. Of those who did jump, 18 are seriously injured.

May 28, 1977—The Beverly Hills Supper Club in Southgate, Kentucky; more than 3,000 have come to spend the evening in one of the club's many banquet rooms or in the Cabaret Room, where singer John Davidson prepares to entertain the capacity crowd of 1,200. Faulty electrical wiring in the wall space on one of the banquet rooms sparks a fire. A busboy goes to the microphone in the Cabaret Room to announce, "There's a small fire. Everyone stay calm and please exit the building." But the crowd stays at their tables, thinking that what they are hearing is just part of the comedy entertainment.

Within moments, a door bursts open and flames engulf the room. Heavy smoke begins pouring through the entire three-story structure, and the electrical system fails, plunging the place into darkness. With no clear idea of how to get out, the people panic, and 165 are killed.

This morbid litany of disasters points up the all too many reasons why going out for an evening's entertainment can be hazardous to your health. Here are some "sins against fire safety" that are committed over and over again in restaurants, nightclubs, and discos:

◊   Overcrowding

◊   Providing too few exits and not marking them clearly (elevators and revolving doors cannot be considered safe fire exits)

◊   Using jerry-rigged, shortcut construction techniques

◊   Using combustible furnishing and decor

◊   Using materials that produce toxic fumes when they burn

◊   Providing no smoke- or heat-detection system nor a sprinkler system (not one of the places described above was equipped with a sprinkler system)

◊   Failing to call the fire department at the first moment a fire is detected

◊   Failing to evacuate patrons at the first sign of trouble

Getting Out—Alive

1. Use common sense in choosing a restaurant, nightclub, or disco. Is the place a rabbit warren of rooms and corridors? Would you have to travel any great distance to an exit? Are there enough clearly marked exits to accommodate the number of people there? Are the routes to the exits clear of all obstructions? Are the exits themselves unlocked and clear of obstructions? Do the doors open outward rather than inward? Does the place pack in more customers than occupancy laws allow? If you cannot satisfac-

torily answer every one of these questions, don't waste your time there. It's too dangerous a risk for the fun involved. Find somewhere else—someplace safe.

2. When you decide on a place, try to be seated near an exit and away from the kitchen. (Close to 40 percent of restaurant fires begin in the kitchen.) Sitting close to an exit will lessen the distance you have to cover in case of emergency. Remember, you may have to operate in the dark, and the fewer overturned tables and chairs you have to clear, the better.

3. When you are seated, make a mental note of all the exits available to you and determine what your path would be to each. Also calculate the distance to each, to help you operate in the dark if that becomes necessary. Remember that the way you entered a room can easily be the first way to be blocked completely by fire. Make a mental note of all exits.

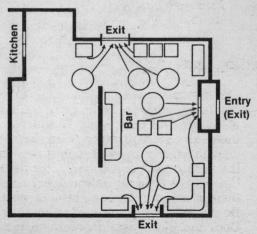

Preferred places to sit in a restaurant

4. Inform the management as soon as you detect any sign of fire and get out in a quick and orderly manner, as you were trained to do in school fire drills.

5. Do not try to retrieve any belongings from the checkroom. JUST GET OUT.

6. Call the fire department as soon as you are outside, if they have not already arrived. It is possible that everyone else thought someone else was turning in the alarm, and as a result, no one did.

SCHOOLS

Perhaps no fire is more pitiable and savage in consequence than a school fire. In this century, three American school disasters have claimed 565 victims, mostly children. Two of the disasters occurred through fires, and one was the result of an explosion. In 1908, fire at the Lake View Elementary School in Collinwood, Ohio, claimed 175 lives. In 1937, an explosion at the London, Texas, Community High School killed 294. And in 1958, 95 young children lost their lives in the fire at Our Lady of Angels Elementary School in Chicago, Illinois.

What can you do to try to keep your own children safe from school fires? There are at least two things.

First, take a good look at the school. How well equipped for fire safety is it? Does it meet stringent standards of fireproof construction, plentiful and enclosed stairwells, and a well-organized and well-rehearsed evacuation procedure? If you have any reason to believe that it lacks any of these things, join with other parents to see that the situation is corrected immediately.

Second, show an interest in fire safety to your children. Help them to understand the value of school fire drills and the need for children to give these drills their full attention and cooperation. Impress on your children that fire drills are not jokes or happy respites from the regimen of the school day. Stress the need for leaving school during a fire drill in an orderly and quick manner. And just as you have worked out a plan of evacuation from your home in case of fire and have sketched a diagram for it, have your children sketch a diagram of the procedure they are supposed to follow during a fire drill. Discuss it with them just as you discussed your home evacuation plan with them.

STORES AND SHOPPING MALLS

Like other public places, stores and shopping malls have the potential for catastrophic fire. The degree of this potential varies widely among shopping places, depending on such differences as local fire codes, age of the buildings, and fire safety systems installed. Even in the same area, one store may have the most up-to-date alarm, sprinkler, and evacuation systems while another offers no such protection.

The two most serious fire-related store disasters in the United States in recent years could not have been averted, even with the best alarm, sprinkler, and evacuation systems because the mayhem they created resulted from explosions. The first occurred on April 5, 1968, in Richmond, Indiana. There, pre-Easter shoppers were patronizing the stores that lined Main Street. At 1:45 P.M. there was a muffled explosion, and the Marting Arms sporting goods store seemed to lift off its foundation. A second explosion tore the store to pieces, sending out a shower of debris in all directions. A great ball of orange flame shot up and several surrounding stores collapsed. In a matter of moments, the blast had killed 41 people and done $2 million in property damage. The culprit? Fire officials suspected improperly stored gunpowder.

The second disaster took place in a series of stores in Eagle Grove, Iowa. There, on February 2, 1973, a natural-gas explosion ripped through the basement of a hardware store, leveling it and two adjoining stores and killing 14 persons.

A third store disaster was initially set off by a low-order explosion of undetermined origin, but most of the human and property damage that ensued was the result of the fire the explosion touched off. The scene was the Younkers Department Store in the Merle Hay Mall in Des Moines, Iowa; the date, November 5, 1978, a Sunday. The store had not yet opened, but there were 22 employees working in the building. Of these 15 were working on the second floor of the two-story structure.

The explosion occurred at 9:30 A.M., and it blew a security guard through the glass front door. The ensuing fire produced heavy smoke and intense heat and enveloped the workers on the second floor before they could reach an exit. Ten of them escaped through a window, from which they climbed to the roof, where they were later rescued. Ten others were not so lucky; they were dead, not from the flames but from smoke inhalation. Had the store been open and full of customers, the tragedy quite possibly would have been record-making.

The Younkers fire points up two of the ways that stores often fail to provide for fire safety. First, the store had no sprinkler system to retard the spread of fire and the buildup of smoke and heat. Second, several of the exits were not clearly marked. Possibly even some employees might not have known the locations, but fire officials judged that customers would definitely have had difficulty finding them.

Stores of just about any kind stock an incredible variety of flammable goods, so once started, fire has plenty to feed on. Smoke and heat can build up fast, and you do not want to be there when they do.

Shopping with Safety

1. If you are like most Americans, you have a large number of shopping places from which to choose. Heretofore, you probably have made your choices on the bases of price, selection, and convenience. Now add another criterion to determine where you will shop—fire safety. Look around the stores you patronize. Do they have smoke detectors? Do they have sprinkler systems? Do they have alarm systems so that you can both warn and be warned of a fire? Are there plenty of exits available in case of fire— enclosed, fireproof stairwells that are clearly marked and easy to reach? (Remember, you cannot count on elevators. They are not acceptable fire exit alternatives.) The next time you enter each of the stores you patronize, look around and then answer each of these questions. If you come up with the wrong answer to any one of them, take your business to someplace that answers them correctly—for safety.

2. As you move around a store or shopping mall, keep aware of where you are in relation to the exits. This is especially important when you go into trying-on or fitting rooms, where you will be somewhat closed off. Obviously, you do not have to become obsessive about this, but out of the corner of your eye, keep track of your possibilities of escape in case you suddenly have to

make use of them. Good fire-safety consciousness will become naturally habit-forming.

3. In case of fire, act quickly and calmly. Head for
the nearest exit, walking toward it purposefully
and at a surefooted pace. If you need to use
stairs, grasp the handrail firmly and hold onto it
as you walk; you will have a much better chance
of staying on your feet should anyone else panic
and race down behind you. Once again, do not
use an elevator as a means of escape. A stalled
elevator can trap you.

THEATERS AND MOVIE HOUSES

Theater and movie house fires mercifully have not
been major claimers of life in recent years, in large
part due to lessons painfully learned in earlier times.

The nation's worst theater fire prior to this century
occurred at the Brooklyn Theatre on December 5,
1876, the nation's Centennial year. That fire began in
the flies during a performance of a popular melo-
drama, *Two Orphans*. In the absence of a fire curtain,
the flames spread quickly out into the audience. Just
about everyone in the orchestra section, on the
ground level, and in the first balcony was able to
escape through the exits, but the story was different
for those in the upper gallery. More than 500 people
had only one flight of stairs available to them to
reach the exits from their high perch. In the inevita-
ble crowding and panic that followed, 295 perished.
However, their deaths did result in fire codes that
mandated multiple exits from all parts of a theater.

When the Iroquois Theater was built in Chicago a
quarter century later, the builders seemed to have
"gone to school" on the Brooklyn Theatre fire. They

installed an asbestos curtain that could be lowered at the first sign of fire onstage. And they included multiple exits that they felt were sufficient to evacuate the 1,700 persons for whom there was seating. But even this planning did not offer the fire safety needed in time of emergency. It did not take into account the overcrowding of the theater that took place during the run of a popular show. Nor did it make allowance for the fact that some of the exits were not clearly marked or were obstructed in some way.

Both of these unplanned factors existed on December 30, 1903, as the leading vaudevillian of the day, Eddie Foy, starred in a holiday performance of *Mr. Bluebeard*. As in the Brooklyn Theatre, fire sparked to life onstage, this time when an arc light blew a fuse and ignited a piece of theatrical gauze. From there it moved on quickly to oil-painted backdrops. The fire curtain was duly lowered, but it caught on a wire, leaving a gap of 12 feet between it and the stage floor. Once again, the flames moved out into the orchestra, and the race for the exits was on. Unfortunately, the frightened patrons soon found themselves up against exit doors that opened inward. As more would-be escapers crushed in behind them, the press of bodies effectively cut off all possibility of opening the doors. The final death count was a staggering 602.

Since the Iroquois Theater fire, rules for theater safety have been made more and more stringent. Theatrical costumes and scenery must be fireproof. Exits must open outward, by means of fast-unlocking "panic bars." And exits must be clearly marked and totally unobstructed.

Nevertheless, fires still do occur in theatrical settings. A fire in the Ringling Brothers and Barnum & Bailey Circus tent in Hartford, Connecticut, in 1944 claimed 168 victims, most of them children. In 1963, a fire at the Indiana State Fairgrounds Coliseum claimed 74. In 1977, a fire in a movie house in Washington, D.C., claimed 9. So it still is clearly necessary to consider fire as a possibility any time you enter a theater or similar setting.

Rules for Theater Safety

1. First of all, know something about the theater you are attending. Often, the playbill that you are given shows a schematic drawing of the theater. Take a few moments to study it to help you relate your location to the exits and where they lead.

2. Look around you. Are there signs of overcrowding? Many theaters have occupancy permission for standees, but the number is strictly limited to a single row of them at the back of the orchestra. If there are more than these few, the theater is overcrowded and fire safety is impaired.

3. As in other places of public assembly, spot the two exits closest to you and plot your route to them. Count the number of seats and rows you will have to pass to get to them to help you gauge the distance.

4. In case of an emergency, the manager generally comes out to make an announcement from the stage. Listen carefully and be cooperative in following any instructions given for evacuation. Remain calm and move toward the exit in an orderly way. Panic is the major enemy here, so resist it and help those around you to resist it, with quiet words and a helping hand, if necessary. This can be a situation in which everyone's individual safety is totally interdependent.

CHAPTER IV
KEEPING SAFE IN THE WORKPLACE

Fire Hazards in the Workplace

Escape Measures

Sprinkler Systems

High-Rise Offices

FIRE HAZARDS IN THE WORKPLACE

There are so many kinds of work being done in this world that it is impossible to spell out the fire hazards connected with each. Special kinds of work have their own special kinds of hazards. For example, work in mining, chemicals, steel, and gas and oil processing and storage has about it the possibility of explosion. Work in wood sanding and refinishing has about it the hazards of highly combustible wood dust and flammable liquids such as polyurethane finish.

Perhaps, then, the best advice is to study your own line of work and your own workplace to determine the specific fire hazards and how they might best be minimized. Also, familiarize yourself with the fire safety rules for your particular workplace (often they are posted around factories, laboratories, and other types of workplace), and obey them. Encourage others to obey them too, for your safety as well as theirs.

However, there are fire hazards common to many workplaces, so you should be ever mindful of the following dangers the next time you go to your own place of work or visit the workplace of another.

Flammable Liquids

These are found in any number of work settings, from the graphics studio where rubber cement is used, to the dry cleaners where cleaning solutions are used, to the printing press room where highly volatile ink thinners are continually used. Follow the rules for using these liquids safely, as presented on pages 43–47.

Cutting, Welding, and Grinding

These operations all can throw off sparks, so be sure that you protect yourself with a protective cover or shield. Take great care that you do any of these things with plenty of clearance from any combustibles that might be ignited by sparks, hot metal, or flame.

Electrical Hazards

Beware of the hazards already spelled out on pages 25–29 regarding electric cords and electrical wiring. These include such hazards as overloading of sockets, frayed cords, exposing wiring, and strange smells or power variations that indicate an electrical problem. If you suspect that anything is wrong with a machine at which you are working, turn it off and have it serviced.

Collected Combustibles (Also known as messy housekeeping)

Keep your work area clear of any clutter, but particularly clutter that can catch fire. Here, neatness definitely does count. Don't allow trash or job waste to build up around your work area. Such clutter presents the dual dangers of catching fire and of impeding your escape from a fire.

Smoking Materials

Obey the rules set up for smoking in your work area. If smoking is prohibited, don't smoke. There is probably a good fire safety reason for the prohibition.

If you can smoke in your work area, follow the commonsense precautions described on pages 73–74 under "Wastebasket Fires."

Do You Wonder About Fire Protection Standards in Your Workplace?

If you are concerned that your workplace does not have adequate fire protection, you may want to obtain a copy of the government standards, put out by OSHA, the Occupational Safety and Health Agency, that concern workplace fire protection. These standards cover such matters as fixed and portable extinguishing systems and equipment, training workers to deal with fires, fire detection systems, and emergency escape plans. You may obtain a copy from the OSHA Office of Publications, Room S1212, U.S. Department of Labor, Washington, D.C. 20210.

ESCAPE MEASURES

So many Americans are working in high-rise buildings these days that a section appears on safety in that special setting on pages 107–108. But whether your workplace is a high-rise or not, there are certain escape measures that you should be able to count on. You should know the answers to these questions:

1. Does your company have a fire emergency plan?

2. Are the exits from your workplace clearly marked and kept free of obstructions at all times?

3. Have you worked out your own personal escape

plan from your usual work area and from other areas in the workplace where you spend time?

4. Do you know what steps to take should fire occur in your workplace?

The section "High-Rise Offices" on pages 107–108 contains information about all of these escape measures. Read it and apply what it has to say to your own work setting.

SPRINKLER SYSTEMS

One of the greatest contributors to fire safety in the workplace is the automatic sprinkler system, so you are fortunate if your workplace is equipped with such a system. If it is not so equipped, you may want to start a campaign to have one installed, for there is no record of a fire that killed two or more people in a building where a sprinkler system was operating and properly maintained. Given this information, one can only wonder why sprinkler systems are not mandatory in *all* buildings.

Simply put, an automatic sprinkler system is composed of a piping system suspended from the ceiling and a series of sprinkler heads attached at intervals to the piping. The pressurized water within the pipes does not surge out of the sprinkler heads because they are kept closed by a heat-sensitive element. (The effect is rather like closing off the nozzle of a hose.) In the event of fire, the heat-sensitive element melts, creating an opening through which water sprays out in all directions. (This effect is rather like opening the nozzle of a hose.) Sprinkler systems operate on the principle of using the fire's own heat to put out the fire.

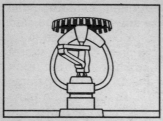

1. Sprinkler head before action of heat

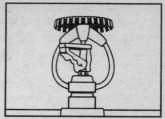

2. Heat starts to melt soldered link

3. Solder completely melted, causing complete separation of link and lever arrangement, and water starts to flow

4. Full force of water strikes deflector causing a spray pattern

Sprinkler head action

Automatic sprinkler systems essentially have four functions:

1. to detect a fire

2. to sound an alarm

3. to extinguish the fire, or at least

4. to prevent it from spreading

Sprinkler systems have a long and quite reliable history, going back to 1723. In that year, one Ambrose Godfrey patented a device later described as a "cask of fire extinguishing liquid containing a pewter chamber of gunpowder. This was connected by a system of fuses which were ignited, exploding the gunpowder and scattering the solution . . . the said vessels so filled and prepared to be made use of by firing the said fuse and then flinging the said vessel into the place where the fire broke out, which upon explosion of the gunpowder, blasts out all the flame and the water or other ingredients against the parts that were on fire, and do damp and suffocate the same so effectively that any man may safely enter the place and with the proper implements may totally extinguish the remaining fire."

In the 2½ centuries since then, sprinkler systems have undergone numerous improvements and refinements, largely in response to the demands of business. Indeed, it is thought that without reliable automatic sprinkler systems, American business, industrial, and mercantile growth would have been severely retarded. Without sprinkler systems, fire damage in the workplace would have been too great and fire insurance too costly.

For the past 80 years, records have been kept concerning the effectiveness of sprinkler systems, and it has been impressive indeed. In the fires studied— 117,770 in sprinklered buildings—the sprinklers proved 95 percent effective in putting out the fires.

The value of having automatic sprinkler systems in the workplace is eminently clear. For one thing, they save lives and prevent injuries. For another, they protect property. And for a third, they prevent the interruption in doing business that severe fires cause.

HIGH-RISE OFFICES

Today, for the first time in history, there are more white-collar workers in the Unitef States than blue-collar workers. And as the number of office personnel grows, so does the number of high-rise office buildings to accommodate them.

High-rise office buildings present a number of challenges to fire safety:

1. They extend far beyond the reach of conventional fire department aerial equipment. (In the case of the Sears Tower in Chicago, 100 stories beyond.)

2. They have the potential for a severe stack effect. They can produce strong updrafts that can quickly draw smoke and flames upward.

3. They can take a long time to evacuate in case of fire, especially when the number of available stairwells is reduced by the fire's effects.

Obviously, if you work in a high-rise office building, it may be worth your life to do all the planning and take all the precautions you can to ensure your safety in a fire emergency. If you don't think your employer or building manager has done enough to preplan the fire safety of your office, lobby as forcefully as you can for a procedure that meets your or your fire department's standards. Or complain to your boss or local fire chief. Don't worry about being considered a crank or a nag on the subject of high-rise fire safety—the stakes are just too great.

Companywide Planning for Fire Safety

Your company itself should have an overall plan for fire safety, organized in concert with the building as a whole. In some areas of the United States fire safety plans for high-rise buildings are mandatory, while they are not in others. In my mind a high-rise is any building that goes beyond the reach of the local fire department's ladders. A fire on the seventh floor is as dangerous as a fire on the eighty-seventh, and the building plan should be the keystone in your own personal plan to protect yourself. Such a plan should include the following:

◊ A well-worked-out evacuation procedure, complete with floor plans, including diagrams of exit stairs and where they lead.

◊ Diagrams posted on the walls showing "You Are Here," making it possible for anyone to see at a glance his or her location in relation to fire exits.

◊ A fire alarm system.

◊ A fire safety director to coordinate fire safety for the entire building.

◊ A fire brigade, made up of wardens from the building service staff who are well trained in evacuation and other fire procedures.

◊ A system of deputy fire wardens, made up of dependable company employees who are thoroughly familiar with the fire exits in their area, with the use of the fire alarm system, and with the evacuation procedure.

◊ A schedule of fire drills in which the director and the wardens, deputy wardens, and employees all practice what each should do in the event of fire.

Your Personal Plan for Fire Safety

1. Take a walk, pad and pencil in hand, around your office floor to familiarize yourself with the location of exits from it. Do the same on any of the other floors to which your job takes you—conference rooms, cafeteria, and so forth. Note

what exits you might use if the fire alarm rang when you were in any of these locations. Also note the location of fire alarms and how to use them.

2. Draw a diagram of your office (where you probably spend most of your time), showing it in relation to at least two exits—the nearest one and an alternative one, should the nearest one be blocked by fire. Note on the diagram how many office doors there are between your office and the exit—this will be invaluable information should you find yourself in darkness either because of electrical failure or heavy smoke.

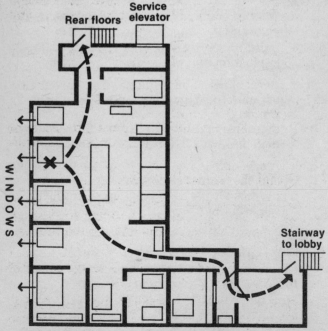

Make a diagram of your office in relation to two exits

Memorize the diagram but keep a copy nearby to refresh your memory from time to time.

3. Keep a flashlight in your desk, again in case of darkness.

4. Familiarize yourself completely with the building evacuation plan and your part in it. And take the fire drills seriously, giving them your full attention and cooperation.

5. Learn if your building has any "refuge areas," often enclosed stairwells, where you should be safe if exits are blocked. Work out your own route to them.

6. See that you always have a clear path from your desk to the exits. Don't ever allow your escape route to be blocked by furniture, boxes, or any other obstacle.

7. Tape the number of the fire department to your telephone.

Taking Action Should Fire Strike

1. If you see any signs of fire, even a little smoke or a small fire, call the fire department immediately.

2. Sound the nearest fire alarm.

3. Get out and encourage your fellow workers to get out too, following the building fire evacuation plan and the instructions of the fire warden.

4. Close doors behind you to hamper the spread of the fire.

5. Go to the exit stairs. DO NOT USE THE ELEVATORS! Elevators can act as firetraps because smoke and flames can shoot up elevator shafts.

6. If there is smoke along your path, crawl rather than walk. Smoke rises, so try to stay beneath it.

7. Do not open any door before feeling it first. If it is hot, do not open it at all. The heat means that there is fire on the other side of it.

8. If you find that you are cut off, don't panic. Get to a room with a window and a phone. Close the door behind you, and seal off the cracks around the door and vents. Open the window at top and bottom and hang something out of it to signal where you are. Also call the fire department to tell them your exact location—floor number and placement on the floor.

In any emergency in a high-rise, remember these basic rules:

DON'T PANIC.

DON'T USE ELEVATORS AS ESCAPE ROUTES.

STAY LOW.

CHAPTER V
KEEPING SAFE WHILE TRAVELING

Airplanes

Camping

Car Fires

Cruise Ships

Hotels and Motels

Portable Smoke Detectors

Portable Smoke Masks

Powerboating

AIRPLANES

Flying in commercial airplanes now is the safest way to travel, but there are enough airplane accidents each year to keep us reminded that airline flight is by no means fail-safe. Many of the fatalities from airplane accidents are caused by impact, and there is nothing you as a passenger can do about that. But many other fatalities and injuries are caused by the fire that often follows an accident, generally suffered on takeoff or landing. In the case of an on-the-ground fire, your calmness, preparation, attention, and cooperation can determine the difference between life and death for you.

Don't Be a "Million-Mile *Macho*"

Perhaps the first lesson to learn has been well stated by Sarah Uzzell-Rindlaub, an expert in air safety procedures and an instructor in airplane survival techniques. (Ms. Uzzell-Rindlaub came by some of her expertise in the most dramatic way— while working as a flight attendant she survived two airplane crash situations in one seven-week period.) She points out that too many passengers—the "million-mile *machos*," as she calls them—simply won't pay attention to the flight briefings that precede each takeoff. According to her, "First- or second-time fliers are more likely to survive. They're more scared or more curious, and they listen. Every plane is different—the exits work differently—and knowing what separates you from an exit could save your life."

Her views are borne out by an airline fire that took place in Connecticut following a faulty landing. Of the twenty-five passengers, only one, a self-admitted "apprehensive flier," had paid strict attention to the flight briefing and had carefully studied the location of exits on the emergency evacuation instructions in the seat pocket in front of him. When the fire broke out, he made his way quickly to a window exit, opened it, and escaped. Only one other passenger had the presence of mind to follow his lead, and she too escaped. All the other passengers, though they were moving around, seemed unable to act or unaware of what action to take. They all perished.

In this case, as in so many others, a little healthy apprehension that led to attention made a life-and-death difference.

Ten Steps Toward Surviving an Airplane Fire

1. Pay close attention to the flight briefing, no matter how many times you've heard it or similar ones before. Make particular note of the locations of all the exits.

2. Study the emergency evacuation instructions provided for you in the seat pocket in front of you. Get a visual sense of where you are on the plane as a whole and your location in relation to the nearest exits. Also acquaint yourself with the procedures for bracing, opening door and window exits, over-the-wing escape, and chute escape.

3. Count the rows of seats between you and each of the nearest exits. You may be operating in the darkness in an emergency, so you may have to rely on your sense of touch rather than sight.

4. Practice fastening and unfastening your seat belt a few times so that you are fully conscious

Count the number of seats to the nearest exit

of how it works and thus can get out of it quickly if necessary. The best place to wear your seat belt is slung around your hips and tight enough so that you have to sit up straight.

5. See that any hand luggage is placed out of the way under the seat ahead of you and that your feet are perfectly free of any obstructions. Speed is all-important in surviving an airplane fire, so you do not want anything around that will impede your movement.

6. If there is any sign of trouble, concentrate single-mindedly on the instructions the flight attendant gives you about bracing. Don't let fear distract you.

7. Do not leave the brace position until the plane comes to a complete stop. There may be more than one impact.

8. If smoke enters the cabin, crawl under the smoke toward the exit, keeping your head at armrest level. Remember that onrushing smoke can easily cause panic, so concentrate on your orientation, and crawl quickly but in an orderly fashion.

9. If you are using a chute escape, take off your shoes first. When you reach the bottom of the chute, move away quickly to avoid being hit by the passenger following you.

10. Move away from the plane as quickly as you can.

CAMPING

Camping has become one of America's fastest-growing leisure-time activities, as the signs "ALL FILLED" at the entrances of some national parks attest. And like other, less bucolic activities, camping can involve fire hazards that must be avoided. Chief among the producers of these hazards are campfires, camp stoves, lanterns, and heaters. Care must be exercised with every one of these, both for the protection of human life and for the protection of the natural setting in which camping takes place.

If you or members of your family are camping enthusiasts, you should recognize the hazards and take these precautions to minimize them:

1. Build a campfire only in a clear area that is a safe distance *downwind* from tents, bedding, people, and any kind of fuel. Avoid any area where there are either tree stumps or overhanging branches. First dig a pit and surround it with a ring of stones. Then build your fire in the pit, safe in the knowledge that it cannot spread.

2. Do not use flammable fuel to start a campfire; use twigs for kindling instead.

3. Keep children away, and never let them start a campfire themselves.

4. Never leave a campfire unattended.

5. When you have finished with the campfire, drench it thoroughly with water. Stir the soaked ashes and smother them with dirt or sand. This procedure will ensure that no potentially dangerous sparks remain.

6. Choose and use tents and sleeping bags that are made of flame-retardant material. Read the labels. Tents or sleeping bags without fire retardants can ignite, and they can burn quickly—with you in them.

7. Never use candles inside tents. Use flashlights instead.

8. Exercise extreme caution when fueling and using camp stoves, lanterns, and heaters. Understand and follow the makers' instructions for their use, and keep them a safe distance away from tent sides and sleeping bags. Take the same precautions with camp heaters that you would with space heaters at home. (See pages 67–70.)

9. Carry fuel for camping equipment only in an aluminum container with a leakproof cap. And be sure that you store it upright in your pack.

10. Let any fueled camping equipment cool down thoroughly before attempting to pack it.

CAR FIRES

Each year more than fifteen hundred Americans die in car fires, nearly one fifth of the annual fire death toll. The great majority of them were trapped in their cars as a result of an accident. But car fires do not have to be accident-involved. Of the more than six hundred thousand motor vehicle fires per year, a great number result from indirect causes such as rubber wearing away from wiring and exposing it, thus touching off flames.

If you are driving along and suddenly smell or see smoke, or if there is any indication that you may have a car fire, approach the problem in a reasonable and orderly manner. Your main concerns should be to save yourself and your passengers, and to avoid endangering others. (Stopping in the middle of the road or leaving the car on an incline without the emergency brake on definitely fall into the "endangering others" category.)

Steps to Take in a Car Fire

1. Pull over to the side of the road.

2. Turn off your ignition.

3. Put on the emergency brake.

4. Open the hood latch. (Many times fire fighters arrive at a car fire and are faced with a locked hood, making it necessary—and often dangerous—to have to force the hood open.)

5. Get out quickly.

6. Open the trunk, if possible.

7. Keep an ABC fire extinguisher in your car. If you think you have a reasonable chance of putting out the fire, use the fire extinguisher on it.

8. Call the fire department. Remember, in a car fire there always is the possibility of a gas tank explosion, and depending on the location of the fire, a car can become a bomb. So stay clear and get professional help.

An Added Precaution

Use great care when you are transporting substances such as propane gas in your car. One Islip, New York, man was using his car to bring home a twenty-pound propane gas tank that was not fitted with a valve guard. During the course of the stop-and-start traffic along the way, the tank tipped over and the valve assembly cracked. As gas seeped into the passenger compartment, it came into contact with a spark from the electrical wiring system. The flash fire that followed set fire to the car upholstery and severely burned the driver. If you have occasion to transport propane gas in your car, see that the tank has a valve guard and that you position the tank so it will not fall over.

Don't transport gasoline in the trunk of your car either. If you must fill your household machine, lawnmower, power saw, or the like, it would be safer to siphon the gasoline from your car than to transport it in a gasoline can in the trunk of your car. A siphoning hose and pump are not very expensive, but remember to have a well-ventilated area to work in, and remember also to use an approved gasoline storage can.

CRUISE SHIPS

Since the TV show *The Love Boat* began its highly popular run in 1977, hundreds of thousands of Americans have made cruise ships the vacation of their choice. Today, some 65 luxury liners regularly call at American ports.

Love Boat's stories probably had prompted many of the luxury liner M.S. *Prinsendam*'s 320 passengers to take to the sea in October 1980. As the gleaming 426-foot ship steamed westward through the Gulf of Alaska, its passengers looked forward to a leisurely month-long cruise to Singapore. "Every day is a day of joy aboard the M.S. *Prinsendam*," a company brochure read. "Six passenger decks are devoted to the luxurious vacation life you expect, with an ambiance of intimacy and charm."

But alarms in the night quickly put an end to their "days of joy." At 1:30 A.M. they were roused from their beds. Fire had broken out, probably the result of a ruptured fuel line that had spurted oil onto hot pipes, causing the oil to burst into flames. For the next 4½ hours the crew fought the flames, but it was a losing battle—the fire was out of control. At 6:00 A.M., the captain ordered, "Abandon ship!"

Drawing on their experience with a practice drill three days earlier, the passengers went to their stations and boarded the lifeboats. Other ships already were steaming to the rescue, and helicopters were overhead. Within 13 hours, all 320 passengers and 200 crew members were saved. Some were suffering from seasickness, the result of being tossed about in their lifeboats, but otherwise they were all alive and well. The fire raged on for days until finally the ravaged hulk slid under the waves and sank to the floor of the gulf.

Planning Your Cruise

Unfortunately, there is no absolutely sure way of ensuring total fire safety aboard ship. The *Prinsendam* had been built only seven years before its sinking, and it conformed to all of the standards of the 1974 International Convention for the Safety of Life at Sea. It also had passed a U.S. Coast Guard safety inspection only five months before the disaster. Yet it was not immune to fire at sea.

Nevertheless, no lives were lost, thanks to the fact that the Prinsendam met and in some ways exceeded ship fire safety standards. If you are planning a cruise, find out about the fire safety standards of the ships you are considering. Heed the warning that one U.S. maritime official has given about some cruise ships currently sailing: "Their paint is fresh and they look nice, but a lot of them are floating firetraps." Be sure you ascertain beforehand that any ship you choose meets at least minimum U.S. fire safety standards. And be sure that the crew is certified as well trained in evacuation procedures.

Protecting Yourself While Aboard

1. Follow the same commonsense rules for fire safety that you would follow in your home or in a hotel, such as not smoking in bed and being careful with smoking materials around upholstered furniture.

2. Familiarize yourself with your surroundings as you would in a hotel. (See pages 126–128.) Study the ship's diagram posted in your cabin or in the corridor to find out where the nearest exits are.

3. Familiarize yourself with the procedures you are expected to follow in case of emergency. Most ships post a list of instructions, written in several languages, in each cabin.

4. Give your full cooperation and attention during the safety and lifeboat drill. International standards demand that such a drill be held at least twenty-four hours after sailing. Don't be blasé about this drill. It is not a lark. Indeed, it can mean the difference between life and death for you.

5. Once you've learned where your boat station is,

practice reaching it from your cabin and from other places on the ship where you are likely to spend time. Work out alternate routes to it as well. This can be fun, and interesting as well, for it will give you a chance to explore the ship more fully.

6. Follow the instructions given to you at your boat station. The crew member who is in charge of each station will demonstrate the procedures you must follow. He or she will explain the use of life vests and will tell you how the lifeboat will be lowered, how you should get into it, and what will happen from then on. Make sure you understand completely, and don't be afraid to ask questions to clarify anything you do not understand.

HOTELS AND MOTELS

Three fatal hotel fires clustered in the three-month period between November 1980 and February 1981—the MGM Grand Hotel in Las Vegas, Stouffer's Inn in Harrison, New York, and the Hilton Hotel in Las Vegas—struck terror into the hearts of Americans who travel. The loss of 119 lives to these hotel fires in that brief time galvanized the nation's attention around the genuine threat to safety that hotel fires create.

Of course, fatal hotel fires are nothing new in American life. Since 1934 alone, there have been at least 130 hotel fires that caused multiple deaths and many more that caused single deaths. Oddly enough, the worst hotel fires seem to come in threes and in a relatively brief period of time. In a six-month period in 1946, the LaSalle Hotel in Chicago, the Canfield

Hotel in Dubuque, Iowa, and the Winecoff Hotel in Atlanta all burned, with a total loss of 199 lives (119 in the Winecoff alone—the nation's worst hotel death toll from fire to date). In a nine-month period in 1963, three hotel fires took 51 lives, and in a similar period in 1970, three more took 67 lives.

Each rash of hotel fires brings on new agitation for stricter and more efficient fire codes, but at best, the degree of safety that fire codes provide is highly uneven. Fire codes for hotels not only vary widely across the country, but they also vary in application in the same area. New codes may not apply to buildings constructed five or ten years earlier. However, there are signs today that many hotels are indeed trying to improve their fire safety procedures, by providing proper and clear escape routes and by informing guests about them.

Nevertheless, a great deal of responsibility for your protection against hotel fires rests squarely with you. Through careful planning, familiarization with your surroundings, reason, and orderly action, you can lessen the chances that your death will become a hotel fire statistic. The first step is easy:

DON'T SMOKE IN BED!

Smoking is the greatest single cause of hotel and motel fires, and most of these fires are bedding-connected.

The second step is just as important:

NEVER USE AN ELEVATOR TO ESCAPE FIRE!

Fire moves upward, and elevator shafts invite its progress. In time of fire, elevators can be traps.

Planning for a Hotel Stay

1. Inquire about a hotel's fire safety before you make a reservation there. It might seem awkward, but you should make a genuine effort to get the reservation clerk to answer some questions. Be insistent. Ask for the manager if you must. Does the motel or hotel have a fire evacuation plan? Does it have a fire alarm system? A smoke detector system? A sprinkler system? And will each of these systems be in operation in the part of the hotel in which you will be spending time? If the answer to any of these questions is no, reserve at a hotel or a motel where all the answers are yes.

2. Current fire equipment generally can reach only as high as the sixth to tenth floors of a building. This fact may make floors at or below this level more attractive to you. Note, though, that helicopters now are used in rescue from highrises. In the MGM Grand Hotel fire, helicopters plucked approximately 250 guests off the roof, guests from upper floors who had been able to reach the roof. In any event, you may want to ask if there is a room available with an adjoining veranda or courtyard or any immediate access to the outside of the building.

3. Carry a survival kit with you when you travel. Components of this kit should include:

   a. A flashlight—lights often go out in a hotel fire; even if they do not, thick smoke can darken corridors.

   b. A portable smoke alarm. (See pages 133–134.)

   c. A portable smoke mask. (This is a new product and should be available soon on the mass market. (See pages 65–66 for additional information.)

4. To develop an understanding of moving about in an unlighted environment, practice exiting a room in the dark at home before leaving for a hotel stay.

Familiarizing Yourself with Your Surroundings

1. When you are taken to your hotel room, ask the bellperson to point out to you the nearby emergency exits. Find out where these exits lead and whether they have signs showing the way out. Also ask for any information the hotel offers on fire evacuation. Many hotels now provide floor plans as well as specific tips for evacuation from the hotel. Also have the bellperson point out the nearby fire alarms and instruct you in their use.

2. After you have checked into your room, go out in the corridor again. Count the number of doors between your room and the nearby exits. If you later find that you must crawl through heavy smoke, you may have to rely on your sense of touch rather than sight. Memorize this information and, as an added precaution, write it down and keep it at hand on your nightstand. Be sure to note any obstruction in your path down the corridor, like an ice machine or a vending machine.

3. Check each nearby exit door to see that it opens. If there is a second exit door to the stairwell, as there sometimes is, make note of it so you will not be surprised if you encounter it in an emergency situation. Also check that there are no obstructions in the stairwell.

4. When you return to your room, study its layout. Be sure that pathways to the door and to the windows are clear of furniture or luggage.

5. Check your windows. Do they open, or are they sealed? How would you break one if that becomes necessary? Look out your windows to get a good picture of where you are in relation to the outside environment. Are there any ledges or setbacks nearby that might help you to escape?

6. Place the components of your survival kit where they belong. The portable smoke alarm may need to be hung on the entry door or on another door, according to directions for its use. The flashlight, your door key, a smoke mask, and notes for your escape plan should be on the nightstand.

Hotel survival kit and portable smoke alarm

7. Practice rolling out of your bed, grabbing your flashlight and key, and crawling toward the door a few times. This is a procedure you may

have to perform immediately upon being awakened, so be thoroughly familiar with it.

Taking Action Should Fire Strike

1. If a fire begins in your room:
   ▶ Report it immediately by dialing 0 on the telephone.
   ▶ If you have any doubts about being able to put out the fire, leave the room immediately, closing the door behind you to keep flames and smoke from going into the corridor.
   ▶ Activate the fire alarm in the corridor to rouse your neighbors.

2. If you are awakened by smoke, a fire alarm, or yelling or a phone call about a fire:
   ▶ Grab your flashlight and key, roll out of bed, and crawl to the door. Be sure you take your key. You may have to retreat to your room, and you must be able to get back in.
   ▶ Don't take time to gather any belongings, no matter how important they seem.
   ▶ Touch the doorknob and door.

3. If the doorknob and door feel cool:
   ▶ Brace your knee against the door and open it slowly to assess the situation. (Bracing the door will allow you to shut it quickly if fire or heavy smoke fills the corridor.)
   ▶ If the coast is clear, go into the corridor, close your door behind you, and move toward the exit. If there is smoke, crawl along the wall that you know will lead you to the exit, counting the doors between you and it if necessary. If that exit is blocked, move to an alternate one. But DON'T USE THE ELEVATOR!
   ▶ Walk down the stairwell, holding the handrail. (People around you may panic, and hold-

ing onto the handrail will help you keep on your feet.) If the smoke grows too dense on the lower levels, reverse your path and walk toward the roof. Once you reach the roof, prop open the door to vent the stairwell. Go to the windward side of the building to avoid any blowing smoke, and await rescue.

4. If the doorknob and door feel hot:
   ▶ Do not open the door—heat means that there is fire in the corridor. But DON'T PANIC! There still are several ways to protect yourself until help comes. In many cases you can stave off disaster for from thirty to forty-five minutes, which allows plenty of time for rescue. Orderly action is the key to survival.
   ▶ If smoke has come into your room, open the window slightly, top and bottom, to vent it. If there is no smoke, don't open the window—there may be smoke outside. Try to avoid breaking a window, if possible—a broken window cannot be reclosed, and broken windows are dangerous. More importantly, the fire might be on a lower floor also, and the rising smoke might enter your room through a broken window.
   ▶ Use the phone to report where you are and what your predicament is. If there is a window that opens, hang out a sheet to show your distress and location.
   ▶ Turn on the bathroom fan, if there is one, to vent any smoke entering the room.
   ▶ Fill the bathtub with water. It will come in handy if rescue is not forthcoming soon. Do the same with the ice bucket.
   ▶ If smoke is seeping in, soak sheets, blankets, towels, and phone books and stuff them into cracks around the door or anyplace else where smoke is coming in.

▶ Tie a wet towel around your face so you can filter out smoke.

▶ Stand a mattress against the door if you can. Prop up the mattress with a dresser, and keep the mattress wet.

▶ Wrap yourself in a wet blanket and stay low, near the window.

▶ Keep as many doors as possible between you and the fire. If necessary, go into the bathroom, closing the door behind you.

▶ If heat continues to increase, use the ice bucket to throw water against the door and walls to cool them. Keep them wet. And let the tub overflow to soak down carpeting and floors.

▶ If there is fire outside the window, pull down the drapes and move anything else combustible away from the window.

▶ Never consider jumping if you are above the third floor, for the chances of surviving the fall are less than they are of surviving the fire.

▶ If breathing becomes difficult and there is no fire, nor heavy smoke, outside the window, put your head outside to get fresh air. (If the window is sealed and you must break it, be sure to clear away broken glass first. And put a folded blanket over the opening to cut down the chance of being cut.) Use a wet blanket to make a tent over yourself as you lean out your head.

In any hotel fire emergency, remember these basic rules:

1. Don't panic.

2. Take your key.

3. Don't use elevators as escape routes.

4. Stay low.

5. Keep fighting—rescue may be only a moment away.

In planning your hotel stay:

1. Ask questions.

2. Know your floor plan.

3. Pack your survival kit.

4. Practice getting around in the dark.

In-the-room guide:

1. Make a mental note of the locations of the exits.

2. Make a mental note of the evacuation plan.

3. Make a mental note of the location of the fire alarm box or station.

4. Count the number of doors to the nearest exit.

5. Check that the exits open.

6. Study your room.

7. Check that the windows open.

8. Arrange your survival kit.

9. Practice your escape once.

10. Prearrange an outside place to meet loved ones or traveling companions.

In case of fire in your room:

1. Report the fire.

2. Get out.

3. Activate the fire alarm.

In case of alarm for fire:

1. Take flashlight, key, and smoke mask.

2. Roll out of bed.

3. Feel doors before opening then.

4. Head for the nearest exit *stairs*.

5. Head for your prearranged meeting place.

If trapped by fire:

1. Use the phone.

2. Turn off the air conditioning or heater fan.

3. Turn on the bathroom vent fan.

4. Hang a distress sheet from a window.

5. Fill the bathtub.

6. Stuff cracks against smoke.

7. Wrap a wet towel around your face.

8. Wrap a wet blanket around you.

9. Soak doors and carpets.

10. Fight the fire.

Foreign words for "exit":	
French	*sortie*
Spanish	*salida*
German	*ausgang*
Italian	*uscita*

PORTABLE SMOKE DETECTORS

The recent development of portable smoke detectors, small and light enough to pack easily in your luggage, provides a valuable fire safety tool. You can take these devices with you anywhere you travel—to hotels or motels, on camping trips, aboard ship. Even if the places where you stay already have smoke detectors, your own portable model offers you added protection.

Like installed smoke detectors, portable models are powered by a nine-volt battery. Some models are designed to be placed on top of a dresser. Others come equipped with a hanging device so you can place them at the top of a door, near the ceiling, where the smoke rises. (Some of these smoke detectors even have a device that acts as a burglar alarm—it sounds if an intruder enters through your doorway.)

Choosing and Using a Portable Smoke Detector

1. Check the packaging on any model you are con-

sidering, to see that it has been approved by a nationally recognized testing laboratory.

2. Read the instructions for installation and proper care *before* you take a portable smoke detector on that first trip.

3. Test the detector weekly. These detectors come equipped with a testing switch.

4. Don't just buy a portable detector and put it away in a drawer. Make it as much a part of your packing routine as your toothbrush.

5. Familiarize yourself with its sound so that if it awakens you, you will know instantly that you are in an emergency situation.

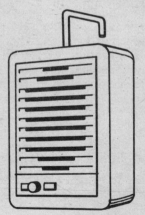

Portable smoke detector

PORTABLE SMOKE MASKS

Smoke masks are a new development and not yet widely in use. Several models of portable smoke masks have been proven to be ineffective in filtering carbon monoxide, so make sure whichever smoke mask you choose to buy is laboratory tested and approved by a recognized organization. Before traveling, practice putting the smoke mask on and off.

POWERBOATING

Powerboating is a craze that has swept the United States since World War II. Today there probably are at least five million powerboats in use here. If you are a powerboat enthusiast or are thinking of becoming one, follow these important rules for fire safety aboard ship.

1. Don't smoke when you are fueling your motor.

2. See that all possible ignition sources—motor, fan, heating device—are turned off before fueling. Remember, gasoline vapors travel.

3. Don't fill the tank to the very top—leave a little room for fuel expansion.

4. Wipe up any gasoline that spills.

5. Wait for any fuel odors to go away before starting the motor—the gasoline vapors must dissipate first.

6. Don't allow oily rags or other combustibles to accumulate in the bilge. Keep the area clear.

7. Keep a Class AB, BC, or ABC fire extinguisher on board and handy, and know how to use it. (See "Fire Extinguishers," pages 32–36.)

INDEX

ABOUT THE AUTHOR

DENNIS SMITH was a New York City fireman for more than seventeen years. He rose to national fame with the publication of his big bestseller, *Report from Engine Co. 82*. He is now publisher and editor-in-chief of *Firehouse*, a monthly magazine for fire fighters with a readership of over 700,000.

Facts at Your Fingertips!

We Deliver!
And So Do These Bestsellers.

☐	23188	**BEVERLY HILLS DIET LIFETIME PLAN** by Judy Mazel	$3.95
☐	22661	**UP THE FAMILY TREE** by Teresa Bloomingdale	$2.95
☐	22701	**I SHOULD HAVE SEEN IT COMING WHEN THE RABBIT DIED** by Teresa Bloomingdale	$2.75
☐	22576	**PATHFINDERS** by Gail Sheehy	$4.50
☐	22585	**THE MINDS OF BILLY MILLIGAN** by Daniel Keyes	$3.95
☐	22981	**SIX WEEKS** by Fred Mustard	$2.95
☐	01428	**ALWAYS A WOMAN** (A Large Format Book)	$9.95
☐	22746	**RED DRAGON** by Thomas Harris	$3.95
☐	20687	**BLACK SUNDAY** by Thomas Harris	$3.50
☐	22685	**THE COSMO REPORT** by Linda Wolfe	$3.95
☐	22736	**A MANY SPLENDORED THING** by Han Suyin	$3.95
☐	20922	**SHADOW OF CAIN** by V. Bugliosi & K. Hurwitz	$3.95
☐	20230	**THE LAST MAFIOSO: The Treacherous World of Jimmy Fratianno**	$3.95
☐	20822	**THE GLITTER DOME** by Joseph Wambaugh	$3.95
☐	20483	**WEALTH AND POVERTY** by George Gilder	$3.95
☐	13101	**THE BOOK OF LISTS #2** by I. Wallace, D. Wallechinsky, A. & S. Wallace	$3.50
☐	05003	**THE GREATEST SUCCESS IN THE WORLD** by Og Mandino (A Large Format Book)	$6.95
☐	20434	**ALL CREATURES GREAT AND SMALL** by James Herriot	$3.95
☐	20621	**THE PILL BOOK** by Dr. Gilbert Simon & Dr. Harold Silverman	$3.95
☐	01352	**THE PEOPLE'S ALMANAC #3** by David Wallechinsky & Irving Wallace (A Large Format Book)	$10.95
☐	20356	**GUINNESS BOOK OF WORLD RECORDS— 20th ed.** by McWhirter	$3.95

Buy them at your local bookstore or use this handy coupon for ordering:

Bantam Books, Inc., Dept. NFB, 414 East Golf Road, Des Plaines, Ill. 60016

Please send me the books I have checked above. I am enclosing $_____ (please add $1.25 to cover postage and handling). Send check or money order —no cash or C.O.D.'s please.

Mr/Mrs/Miss_____

Address_____

City_____ State/Zip_____

NFB—2/83
Please allow four to six weeks for delivery. This offer expires 8/83.

DON'T MISS
THESE CURRENT
Bantam Bestsellers